Healthy Ageing and Longevity

Volume 20

Rapidly changing demographics worldwide towards increased proportion of the elderly in the population and increased life-expectancy have brought the issues, such as "why we grow old", "how we grow old", "how long can we live", "how to maintain health", "how to prevent and treat diseases in old age", "what are the future perspectives for healthy ageing and longevity" and so on, in the centre stage of scientific, social, political, and economic arena. Although the descriptive aspects of ageing are now well established at the level of species, populations, individuals, and within an individual at the tissue, cell and molecular levels, the implications of such detailed understanding with respect to the aim of achieving healthy ageing and longevity are ever-changing and challenging issues. This continuing success of gerontology, and especially of biogerontology, is attracting the attention of both the well established academicians and the younger generation of students and researchers in biology, medicine, bioinformatics, bioeconomy, sports science, and nutritional sciences, along with sociologists, psychologists, politicians, public health experts, and health-care industry including cosmeceutical-, food-, and lifestyle-industry. Books in this series will cover the topics related to the issues of healthy ageing and longevity. This series will provide not only the exhaustive reviews of the established body of knowledge, but also will give a critical evaluation of the ongoing research and development with respect to theoretical and evidence-based practical and ethical aspects of interventions towards maintaining, recovering and enhancing health and longevity.

Khachik K. Muradian · Vadim E. Fraifeld

Rejuvenation and Longevity

Introduction to Rejuvenology

Springer

Khachik K. Muradian
Department of Biology of Aging and Life Span Extension
State Institute of Gerontology
Academy of Medical Sciences of Ukraine
Kiev, Ukraine

Vadim E. Fraifeld
The Shraga Segal Department of Microbiology
Immunology and Genetics
Center for Multidisciplinary Research on Aging
Ben-Gurion University of the Negev
Beer-Sheva, Israel

ISSN 2199-9007 ISSN 2199-9015 (electronic)
Healthy Ageing and Longevity
ISBN 978-3-031-64994-3 ISBN 978-3-031-64995-0 (eBook)
https://doi.org/10.1007/978-3-031-64995-0

This Springer imprint is published by the registered company Springer Nature Switzerland AG
The registered company address is: Gewerbestrasse 11, 6330 Cham, Switzerland

Preface

"Since the beginning of time", rejuvenation is a human aspiration. How many heroes of fairy-tales and real folk explorers wandered in vain through forests and other inaccessible places in search for the elixir of youth? Their expected failure made rejuvenation a symbol of sorcerers, alchemists or up to charlatans. Like the discouraging disputes on perpetuum mobile, rejuvenation was confined from serious scientific discussion as an impossible abstraction. Even now, most scientists and laymen are skeptical about reality of rejuvenation. Being largely expelled from the academic milieu, the studies on rejuvenation acquired numerous limitations. Preparing a book in these obstacles was a real challenge, and we are sorry for the objective and subjective shortcomings and mistakes that could hardly be avoided. Nevertheless, during the last decade, rejuvenation has acquired a realistic shape and becomes a subject of scientific analysis. It would not be an exaggeration to say that right now a new science is emerging—*Rejuvenology.*

Various approaches and concepts have been proposed and tested up to date. Yet, there are neither basic theories or hypotheses of rejuvenation nor recognized definitions, hallmarks, etc. All this definitely requires a comprehensive critical analysis. This book is the first of its kind, in which the accumulated data on rejuvenation is systemized and analyzed, starting from the evolutionary aspects of the origin of life, rejuvenation and immortality to the evidence indicating that biological time could be reversible. Despite the obvious progress in the field, scientific and near-scientific community is still divided into two "opposing camps"—for and against rejuvenation and radical life span extension. The book can be interesting for both parties and could hopefully be useful for the establishment of mutually beneficial compromises. Thus, we hope that our work could serve as a basis for further creative discussion of the problem and will stimulate research in the field.

We apologize to all the authors whose contributions we could not cite owing to space limitations. We thank our colleagues and students for fruitful discussions and their real contribution to the field. We are particularly grateful to Dr. Ekaterina Rudnitsky for her assistance in preparing this book.

Kiev, Ukraine Khachik K. Muradian
Beer-Sheva, Israel Vadim E. Fraifeld

Contents

Chapter 1
Introduction to Optimistic Gerontology

Abstract Aging, rejuvenation and immortality are the apexes of a triangle full of amazing paradoxes and gems of simple truths. Their discussion often provokes heated confrontation and disputes warmed up by a wealth of fascinating and yet controversial investigations.

Keywords RNA-world · FUCA · Definitions of aging · Rejuvenation · Immortality

Immortality is for the gods, but man's search for its elusive secrets, perhaps as old as man

himself, will continue.

—Wan Po

1.1 Paradoxical Triangle

Aging, rejuvenation and immortality are the apexes of a triangle full of amazing paradoxes and gems of simple truths. Their discussion often provokes heated confrontation and disputes warmed up by a wealth of fascinating and yet controversial investigations (Frolkis and Muradian 1991; Abramovich et al. 2008; Lewis and de Grey 2024; Yücel and Gladyshev 2024). Is aging obligatory for all biological objects? Can rejuvenation counterbalance the negative effects of aging? Is immortality achievable? Rapidly accumulating data allows for shedding some light on these and related sophisticated issues. In fact, this book is a first attempt to systemically analyze what rejuvenation is and its relationships with aging, longevity, origin of life, and the existence of life as is. In this introductory chapter, we describe in brief the main problems, hypotheses and questions that are discussed in the book.

K. K. Muradian and V. E. Fraifeld, *Rejuvenation and Longevity*, Healthy Ageing and Longevity 20, https://doi.org/10.1007/978-3-031-64995-0_1

1.2 RNA-World and FUCA

Conversion of chemical substances into a self-replicating living system is the most fateful event and thrilling evolutionarily transformation. This event manifested the transition from the 'cold chemistry' to the 'warm biology'. According to the RNA-world concept, biological life has been raised due to the unique universality of RNA molecules, carrying simultaneously genetic, catalytic, structural, and regulatory functions. RNA can store and replicate genetic information as DNA and at the time catalyze biochemical reactions critical for life as protein enzymes. Moreover, two RNA molecules can indefinitely long catalyze each other's synthesis in a self-sustained and exponentially amplifying manner in the absence of DNA, proteins or other biomolecules. In the reaction, RNA vesicles can increase in size due to the synthesis of new molecules, fragment apart into smaller ones and again regrow into new larger vesicles. It is a simple and reliable model of the breathtaking event—shifting of an inanimate matter into an immortal living matter.

The Last Universal Common Ancestor (LUCA) is a widely accepted though still hypothesized concept. It is believed that LUCA gave rise to the three main domains of life (Bacteria, Archaea, and Eukarya). Here, we propose to honor the first two molecules of RNA, from which life has originated, by the abbreviation FUCA (the First Universal Common Ancestor) (Chap. 2).

All in all, it seems plausible that life and rejuvenation originated simultaneously in a self-replicating RNA, thus showing that life and rejuvenation are inseparable. Aging appeared later as an adaptive mechanism for elimination of defective prokaryotes from the population. Rejuvenation has evolved first and foremost and only afterwards appeared aging.

1.3 Concise Definition of Aging and Rejuvenation

There are hundreds of definitions of aging. A concise 'three-words' definition argues that aging is a ***permanent global catastrophe***. In fatality, aging cannot be compared to anything. No cataclysm, war or accident brings as much grief and loss as aging. Millions of people, among them our friends, die yearly because of age-related diseases. A vibrant adult man turns into a decrepit old patient during few decades and passes away often in physical and psychological torments. Therefore, all means are justified to ameliorate consequences of aging.

Rejuvenation is intuitively understood as an ***antipode*** of aging. However, there is no strict mechanistic definition of rejuvenation. It is urgently needed. As a first approximation, rejuvenation could be defined phenomenologically as a process that opposes aging and can decrease the biological age. It could occur at any age, and may enhance the functional capacity, resistance to stress, and reverse, fully or partially, the development of age-related diseases. Although rejuvenation is primarily an organismal feature, it is also relevant to the tissue or cellular level, or even to single biomolecules. It should be emphasized again that rejuvenation is a real natural mechanism for ridding off the load of aging. The major rejuvenative events in multicellular organisms are primarily associated with gametogenesis and early embryo development (Kerepesi et al. 2021; Kerepesi and Gladyshev 2023) (Chaps. 4, 7, 8).

1.4 Aging and Rejuvenation: Cooperation Instead of Confrontation

All hypotheses of aging claim that the avalanche of age-related changes is so destructive and pervasive that a cell will inevitably accumulate lethal damage and die. Yet, the rate of damage could be slowed down close to zero as it occurred in self-replicating RNAs and further in protocells and prokaryotes. *Rejuvenation paradigm* posits claims that modern biological life originated due to intensively replicating immortal protocells and prokaryotes. At a certain rate of a cell division, dilution of damage can reach the level when a cell has 'neither possibility, nor time for aging'. The immortality program could be fixed in a cell genome and transferred to the next generations, including cells of multicellular organisms. Since then, biological life has been thriving for billions of years due to the efficient defense and rejuvenative systems. Aging and rejuvenation are tightly interconnected and well-balanced. The question is how to shift this balance towards rejuvenation and whether we have any tools in hand for doing so.

1.5 Transposons, Aging, and Rejuvenation

Transposable elements (TEs) are relics of ancient viral invasions. With few exceptions, TEs are discovered in all sequenced species and tissues of metazoans, with the highest level of expression in ESCs and reproductive organs. TEs could be symbiotic in embryogenesis and fetal development, being significant for embryonic rejuvenation, silent in adulthood and loosely controlled and harmful in aging. They could be key players in the zygotic genome activation (ZGA). Indeed, ZGA starts from the heterochromatin relaxation followed by the burst of transcription activity. The mobility of endogenous retroviruses (ERVs) and their multiple positioning could give them advantages in opening the relaxation sites.

It is hardly believable but around 70% of the human genome is derived from retroviruses. During the last 10 million years, integration of ERVs in hominoid genomes decreased by several fold. Humans demonstrated especially strong protection against new viral insertions. As we have shown, mammalian longevity correlates with the ability to resist the integration of new EVRs (Chap. 7). Whatever the case, it would be hardly possible to discuss human rejuvenation and longevity without considering the putative role of TEs in embryogenesis and aging.

1.6 Is Immortality Possible?

Infinity is a philosophical or mathematical abstraction. In the real physical world, all events have an arbitrarily large or small but finite probability of realization. Therefore, immortality cannot exist in the real world. Limited life expectancy will remain even if all age-related diseases become curable. Of note, in this case the mean human life span could be around the biblical Methusael's 800–1000 years.

Humans are the longest-lived terrestrial mammalian species. Nevertheless, from ancient times humans strived for a longer life and even dreamed of being immortal. As Wan Po elegantly noted, "Immortality is for the gods, but man's search for its elusive secrets, perhaps as old as man himself, will continue." The existence of animals exceeding human longevity by hundred and thousand times supports the idea of 'practical immortality'.

Maximal life span (MLS) of ocean sponges reaches 10–15 thousand years. Another basic metazoan, corals, are the second longest-lived animal species (over 4200 years). Sponges and corals exhibit the highest regenerative capacity among metazoans, which is largely associated with their *stem cell* systems. Even the moderately long-lived species, such as ocean quahog (over 500 years), Greenland shark (over 400 years), tubeworm (over 300 years) or bowhead whale (over 200 years) could be robust samples of longevity, worth of meticulous examination and imitation (see Chaps. 5 and 6).

1.7 Other Sophisticated Issues and Still Unexplored Possibilities

The phenomenon of cellular reprogramming suggests a high plasticity of the cell which 'remembers' the way to youthfulness and which rejuvenation program could be activated by relatively simple means, including external signals (Muradian and Fraifeld 2020; Knyazer et al. 2021; Yang et al. 2023) (Chap. 11). Another feasible perspective for rejuvenation and radical health span extension could be attributed to a proper maintenance and functionality of *somatic pluripotent stem cells* (e.g., Muse, multilineage-differentiating stress-enduring stem cells) (Dezawa 2018) (Chap. 9).

Any rejuvenative interventions in an advanced age would be hardly possible without strong repression of TEs/ERVs expression and mobility. With this in mind, nucleoside reverse transcriptase inhibitors could be included in 'rejuvenative cocktails' as promising components (Brochard et al. 2023). Whatever, multiple targeting seems to be more efficient (Lewis and de Grey 2024). Already successfully tested blood plasma exchange (Mehdipour et al. 2020; Kim et al. 2022) and thus far only suggested by us 'wash-up' technics for dilution and clearance of intra- and intercellular garbage indicate that rejuvenative procedures could be relatively simple and, anyway, much less sophisticated than their results (☺) (Chap. 11).

It is believed that there have been times when the Earth was inhabited mainly by immortal prokaryotes, viruses and unicellular species. At the dawn of multicellularity, symbiosis of these entities resulted in long-lived metazoans which have later been replaced by the current aggressive and relatively short-lived multicellular species. Evolution is a good tactician but not a strategist. Now a man has taken the initiative. Let's see what we can and will do.

Ethics All the materials and data described in the book have been approved by Ethical Committee for Animal Experimentation or Helsinki Committee for Research on Humans of the organizations, where corresponding investigations have been done.

References

Abramovich A, Muradian KK, Fraifeld VE (2008) Have we reached the point for in vivo rejuvenation? Rejuvenation Res 11(2):489–492. https://doi.org/10.1089/rej.2008.0658. PMID: 18442329

Brochard T, McIntyre RL, Houtkooper RH, Seluanov A, Gorbunova V, Janssens GE (2023) Repurposing nucleoside reverse transcriptase inhibitors (NRTIs) to slow aging. Ageing Res Rev 92:102132. https://doi.org/10.1016/j.arr.2023.102132. PMID: 37984625

Dezawa M (ed) Muse cells: endogenous reparative pluripotent stem cells (Advances in experimental medicine and biology, 1103), Springer, 1st ed., 315 pages

Frolkis VV, Muradian KK (1991) Life span prolongation. CRC Press, Boca Raton, p 425

Kerepesi C, Gladyshev VN (2023) Intersection clock reveals a rejuvenation event during human embryogenesis. Aging Cell 22(10):e13922. https://doi.org/10.1111/acel.13922. PMID: 37786333

Kerepesi C, Zhang B, Lee SG, Trapp A, Gladyshev VN (2021) Epigenetic clocks reveal a rejuvenation event during embryogenesis followed by aging. Sci Adv 7(26):eabg6082. https://doi.org/10.1126/sciadv.abg6082. PMID: 34172448

Knyazer A, Bunu G, Toren D, Mracica TB, Segev Y, Wolfson M, Muradian KK, Tacutu R, Fraifeld VE (2021) Small molecules for cell reprogramming: a systems biology analysis. Aging (Albany NY) 13(24):25739–25762. https://doi.org/10.18632/aging.203791. PMID: 34919532

Kim D, Kiprov DD, Luellen C, Lieb M, Liu C, Watanabe E, Mei X, Cassaleto K, Kramer J, Conboy MJ, Conboy IM (2022) Old plasma dilution reduces human biological age: a clinical study. Geroscience 44(6):2701–2720. https://doi.org/10.1007/s11357-022-00645-w.Erratum.In:Geroscience.2023. PMID: 35999337

Lewis CJ, de Grey AD (2024) Combining rejuvenation interventions in rodents: a milestone in biomedical gerontology whose time has come. Expert Opin Ther Targets 18:1–11. https://doi.org/10.1080/14728222.2024.2330425. PMID: 38477630

Mehdipour M, Skinner C, Wong N, Lieb M, Liu C, Etienne J, Kato C, Kiprov D, Conboy MJ, Conboy IM (2020) Rejuvenation of three germ layers tissues by exchanging old blood plasma with saline-albumin. Aging (Albany NY) 12(10):8790–8819. https://doi.org/10.18632/aging.103418. PMID: 32474458

Muradian KK, Fraifeld VE (2020) iPSCs—induced cellular reprogramming. In: Rattan S (ed) Encyclopedia of biomedical gerontology, vol 2. Elsevier, pp 291–298

Yang JH, Petty CA, Dixon-McDougall T, Lopez MV, Tyshkovskiy A, Maybury-Lewis S, Tian X, Ibrahim N, Chen Z, Griffin PT, Arnold M, Li J, Martinez OA, Behn A, Rogers-Hammond R, Angeli S, Gladyshev VN, Sinclair DA (2023) Chemically induced reprogramming to reverse cellular aging. Aging (albany NY). 15(13):5966–5989. https://doi.org/10.18632/aging.204896. PMID: 37437248

Yücel AD, Gladyshev VN (2024) The long and winding road of reprogramming-induced rejuvenation. Nat Commun 15(1):1941. https://doi.org/10.1038/s41467-024-46020-5. PMID: 38431638

Chapter 2
The Origin of Life, Immortality, and Aging

Abstract Biological life and immortality have originated due to the unique ability of RNA molecules to catalyze each other's synthesis in RNA-templated RNA-polymerization reactions. Covered by lipid envelopes, RNA vesicles increase in size through the synthesis of new molecules. The larger vesicles fall occasionally apart into smaller ones and regrow into newer large vesicles. It is a simple and reliable model of self-replicating life origin. By analogy with LUCA (the last universal common ancestor), we propose to honor the first two molecules of RNA from which life originated by the abbreviation FUCA (the first universal common ancestor). According to the well-known dogma, prokaryotes and some unicellular eukaryotes in the permissive conditions generate genetically identical daughter cells without any alterations that can be interpreted as aging. Moreover, they have decreased the population doubling time up to few hours. At these rates of cell division, they have 'neither time nor possibility for normal aging'. Aging has obviously emerged later as a useful tool for quick removal of cellular damage. Altruistic 'kamikaze' cells selectively accumulated cellular garbage, aged and were eliminated by asymmetric partitioning. In the evolutionary odyssey of multicellular organisms, however, aging has changed the role of the useful tool to the most destructive factor. Why has this mysterious transformation occurred and what could be done to recover the reputation of aging in multicellular organisms? In a human body, there are nearly equal numbers of bacterial and multicellular cells. Modifications of immortal 'personalized' cells (e.g., bacteria with the adult stem cells) could be promising. Remarkably, immortal cells strongly increase expression of the old (bacterial) genes and decrease transcription of the multicellular (new) genes. The main difference here is that bacterial cells are immortal in the normal conditions and could age only in stressful environments, whereas the cells of multicellular organisms undergo aging in the normal conditions and could become immortal after 'struggle' with the differentiation factors. This is a short scenario about the origin of life, immortality, and aging.

Keywords The origin of life · Rejuvenation origin · The origin of aging · RNA concept of life · LUCA · FUCA · Senescence · Immortality · Symmetric partitioning · Asymmetric partitioning

K. K. Muradian and V. E. Fraifeld, *Rejuvenation and Longevity*, Healthy Ageing and Longevity 20, https://doi.org/10.1007/978-3-031-64995-0_2

There is nothing inherent in life that implies death.

—August Weissman, 1891 (in Larocca et al. 2021)

2.1 The Origin of Life

The origin of life is a problem almost as complex as life itself is. It is an intricate biological and philosophical issue full of fascinating findings and unsolved questions. To begin with, there are no commonly accepted definitions of life and its origin. Even the simplest living beings are enormously complex, and it is virtually impossible to follow explicit roots of their origin. However, it is generally assumed that biological life miraculously originated only once, and thereafter all extinct or existing life forms have been derived from the first ancestorial prototype of life.

Prior to life's origin, in the prebiotic periods, occurred the activation of inorganic and organic reactions, resulted in generation of the macromolecular precursors, mostly with participation of chemical elements common in biology—carbon, hydrogen, nitrogen, oxygen, phosphorus, and sulfur (Preiner et al. 2020). The emergence of the self-replicating living systems is the most fateful and thrilling transformation in evolution of life, manifesting the transition from chemistry to biology. Several hypotheses of life origin have been developed, among them the prebiotic Soup, hydrothermal vents, hot spring, and RNA worlds are the most known (Damer and Deamer 2020).

2.2 RNA Concept of the Life and Rejuvenation Origin

We prioritize the RNA role in the origin of life and rejuvenation because it is better argued and let see the light at the end of the tunnel. RNA is a simple linear polymer consisting of four major ribonucleotides. Yet, it performs a remarkable diversity of functions in a cell. Recent discoveries of new RNA functions showed that assessment of the RNA as an intermediatory agent in the chain DNA → RNA → protein is too narrow (Breaker and Joyce 2014). There are at least two reasons why RNA could be a prime founder of biological life. Firstly, in the early Earth abiotic environment, RNA obviously was the most available macromolecule because all four ribonucleotides could be synthesized from simple and available inorganic compounds—water and cyanide (HCN) (Saladino et al. 2012). Secondly, in abiotic periods, it was important for the very origin of life whether a molecule can simultaneously carry out several functions. Versatility could be a key issue for developing in various metabolic and regulatory pathways. Above any doubt, RNA is one of the most universal biological compounds, carrying out functions of genetic, catalytic, structural, and regulatory agents (Nudler and Mironov 2004; Lilley and Sutherland 2011). It can store and replicate genetic information as DNA and simultaneously catalyze biochemical reactions critical for life as protein enzymes do (Johnston et al. 2001).

However, the most unexpected finding here is that RNA molecules can indefinitely long catalyze each other's synthesis in an RNA-templated RNA-polymerization reaction. The reaction occurs in a self-sustained and exponentially amplifying manner in the absence of DNA, proteins or other biomolecules (Lincoln and Joyce 2009; Joyce 2009; Joyce and Szostak 2018). RNAs covered by a lipid envelope and supplied by ribonucleotide triphosphates, i.e., just two molecules of RNA, could endlessly multiply and become the ancestor of a living system. We suggested that the encapsulated RNA vesicles would grow in size due to synthesis of new molecules of RNA until the surface tension of the lipid cover will become insufficient to maintain the weight and the structure will fall apart generating smaller vesicles. The new vesicles will do the same—grow and multiply. It is a simple and reliable model of the self-replicating life origin. It seems hardly believable that a task of supernatural complexity, like the life origin, could be resolved so simply. Due to combination of the matrix and ribozyme properties, two molecules of RNA could give birth to the world of self-replicating proto-proto-protocells. It means the beginning of biological life!

In this context, it is important to stress that *biological life and rejuvenation have originated simultaneously* in a self-replicating RNA vesicle. It is one of the principal and oldest biological relationships. In the long run, life, rejuvenation, and immortality remain inseparable. That being said, where the biological life is—look for rejuvenation and immortality.

2.3 LUCA and FUCA

The unique combination of the template and ribozyme properties of RNA ensured creation of the first prototype of the ever-young and potentially immortal protocells which could self-sustain indefinitely long, divide, and give rise to daughter protocells. By analogy with LUCA (the Last Universal Common Ancestor) and for ease of memorization, *we propose to honor the first two molecules of RNA, from which all biological life has originated, by the abbreviation FUCA (the First Universal Common Ancestor).* Life origin through the self-replicating RNAs seems an unexpectedly simple but realistic act of conversion of the 'cold chemistry' into 'warm biology'. We would like to stress again that it also means that life and immortality emerged simultaneously with direct participation and under guidance of RNA.

Further evolution of biological life in the prokaryotic world and beyond seems more predictable (Kunnev 2020). Replication rate has been the prime-important issue for the survival of unicellulars. In the tough struggle for dominance, prokaryotes achieved tremendous successes, e.g., decreasing the population doubling time up to few hours. Combining data on the fossil material and molecular clock analyses, it was estimated that LUCA emerged ca. 3.9 Ga ago, *Eubacteria* and *Archaebacteria* ca. 3.4 Ga ago, and modern eukaryotes ca. 1.84 Ga ago (Betts et al. 2018). Yet, we can only suggest the time of the FUCA emergence. According to the measurements of the carbon-13 inclusions, the earliest terrestrial biosphere has emerged ca. 4.10 ± 0.01 Ga. It is in line with the 4.4–4.0 Ga shown by Detrital zircons from Jack Hills (Western Australia) (Bell et al. 2015). We presume that the FUCA's emerging time could be around 4.3–4.2 Ga ago.

2.4 Senescence and Immortality

Prokaryotes and certain eukaryotic unicellulars, e.g., *Escherichia coli (E. coli)* or fission yeast *Schizosaccharomyces pombe (S. pombe)*, alternate replicative immortality and aging. They solve the alternative between death and life by simple dilution of the cellular damage or asymmetric partitioning. In these unicellular objects, the transition from mortality to immortality can directly be observed at the molecular level in a single transparent cell (Proenca et al. 2018; Steiner 2021).

Living beings are incredibly complex and inevitably accumulate damage. However, it does not mean that aging is also inevitable because there are numerous mechanisms for damage elimination and neutralization. In addition to the complex set of defense and repair systems, *there are two mechanistically simple but highly efficient tools, widely used from the times of life origin: (i) damage dilution due to a cell division, (ii) and excretion of unwanted substances through release of extracellular vesicles* (Stewart et al. 2005; Sheldrake 2022). By the way, eukaryotic cells can dilute damage not only by binary fission but also in multiple fission mode through which the cells undergo several rounds of continuous cell divisions producing up to 16 and more daughter cells without intervening the growth (Jong et al. 2021). They widely utilize dilution and extracellular vesicles for damage cleaning. For instance, it is well-known that streptomycin produces mistranslated proteins vulnerable to oxidation and aggregation. In such stressful conditions, bacteria generate minicells which cannot proliferate because they lack chromosomal DNA. However, minicells excrete the protein aggregates which often accumulate in cells' poles, thus rejuvenating and escaping cells from imminent aging and death. Survival or growth rates were higher in strains with elevated rates of minicell production in response to the antibiotic stress (Rang et al. 2018).

2.5 Symmetric and Asymmetric Partitioning

According to the known dogma, binary fission of prokaryotes in permissive conditions results in generation of two genetically identical daughter cells, without any alterations that could be interpreted as aging (Eswara and Ramamurthi 2017; Nakaoka and Wakamoto 2017; Rang et al. 2018; Sheldrake 2022). However, unicellulars also evolved asymmetric partitioning when two daughter cells differ in elements inherited from the mother cell because of the uneven segregation. The asymmetry means that one part of the progeny receives old constituents of the mother cell, while the other daughter cell produces these elements de novo. As a result, lineages that have to synthesize new elements become senescence-resistant. The cells that inherit old constituents (modified DNA, proteins, etc.) express fewer genes, become prone to slower growth, and eventually exhibit elevated probability of death and aging (Stewart et al. 2005; Ackermann et al. 2007; Proenca et al. 2019; Shi et al. 2020; Pen and Flatt 2021; Steiner 2021). However, life of such aging bacteria could continue over 200 replication rounds. More importantly, a significant part of aging cells can rejuvenate and escape death if returned into normal conditions or the damaged constituents are excreted in the form of minicell vesicles (Wang et al. 2010; Rang et al. 2018). In cases when damage accumulation is autocatalytic and its rate exceeds the effects of dilution and excretion, a cell is doomed to age and die (Hirsch 1978). Thus, bacteria and some unicellular eukaryotes (*E. coli* or fission yeast *S. pombe*) do not age and are usually immortal in normal conditions. However, they can activate aging phenotype when necessary (Winkler et al. 2010).

2.5.1 *Escherichia coli*

The well-known rod-shaped bacteria *E. coli* is a good example of easy switching from replicative immortality to mortality and back. Lineages of *E. coli* exhibit both aging and non-aging phenotypes, depending on the rates of damage accumulation and elimination. At low damage accumulation rates, immortality is dominating in all strains. At the higher damage rates, mutant lineages become mortal but normal wild type cells remain immortal. Data obtained clearly indicates that replicative immortality is the dominating life mode of bacteria in permissive conditions. Damage accumulation rate can become deterministic in selection of the alternative between aging and immortality (Proenca et al. 2019; Steiner 2021).

2.5.2 *Schizosaccharomyces pombe*

Virtually the same kind of relations have been found for the symmetrically dividing metazoan unicellular fission yeast *Schizosaccharomyces pombe* (*S. pombe*). None of the cell lineages grown in favorable conditions showed the signs of aging, primarily manifested in an increased division time and probability to die. In normal conditions, cell death is random and does not follow exponential growth or other type of the mortality increase. However, asymmetrically partitioning cells that undergo stress and inherit large aggregates of damaged proteins do age. Indeed, the sister cells of *S. pombe* which inherit negligible amount of aggregates exhibit no aging phenotype. In contrast, the daughter cell which inherited the protein aggregates replicates slower,

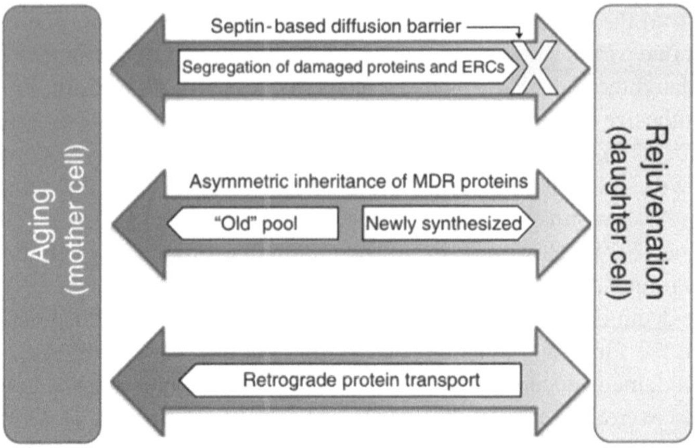

Fig. 2.1 Schematic representation of the polar processes involved in both aging and rejuvenation in the budding yeast *S. cerevisiae* (from Budovsky et al. 2011)

ages and dies, while the other one remains clean and replicatively immortal (Coelho et al. 2013, 2014).

Another yeast species, *Saccharomyces cerevisiae* (*S. cerevisiae*), represents an example of partial rejuvenation. The asymmetric division of budding *S. cerevisiae* ensures the partial rejuvenation of the daughter cell at the expense of the mother cell which retains diverse pro-aging factors (Budovsky et al. 2011). Yet, the daughter cells of *S. cerevisiae* from old mothers display a reduced life span (Kennedy et al. 1994).

As seen in Fig. 2.1, a special retrograde transport ensures removal of the damaged proteins from the daughter cell. This, along with polar protein segregation (including the MDR transporters), polar mRNA transport (Taranukha et al. 2012), and septin-based diffusion barrier, contributes to rejuvenation (at least in part) of the daughter cell.

2.6 The Origin of Aging

Thus, *symmetrically partitioning bacteria and some metazoans exhibit unlimited replicative capacity in permissive conditions.* In stressful situations, however, damage accumulates, thus activating the program of asymmetric partitioning and aging. The aging of such cells could be a useful tool for rapid cleaning up a population from damage. So-called 'kamikaze' cells accumulate most destructive elements by asymmetric partitioning, age and are eliminated with the damage. It is important that aging phenotype develops by asymmetric partitioning only in a limited group of cells. Due to the tremendous replicative capacity of bacteria, these cells could quickly be recovered. The possibility for rapid cleaning up a population from consequences of stress gives certain advantages in competition with other populations which failed to develop such protective mechanism. That said, *aging originated as a repair process for the defense and prosperity of the population.* It occurs only in a limited number of altruistic kamikaze cells ready to pay for the population security.

2.7 Growth and Damage Dilution Are Primordial Mechanisms of Longevity in Multicellulars

The role and efficiency of damage dilution due to a cell division and growth have not been properly appreciated in the biology of aging. Many continuous growers among the animals and perennial plant species which preserve replicative activity and grow during the whole life exhibit no or only negligible aging. Some angiosperm plants demonstrate even negative senescence and rejuvenate, improving physiological performance with age (Munné-Bosch 2015).

The classical evolutionary concepts by Peter B. Medavar, George C. Williams, William D. Hamilton and many others predict the inevitable increase in mortality with the growth rate decrease after maturity. Comparison of standardized patterns of mortality of over 300 vertebrate and invertebrate species has revealed substantial variability in mortality rates, including increased, constant, decreased, humped and bowed trajectories for both long- and short-lived species. Despite that diversity, there was a strong positive correlation between the growth period and longevity (Jones et al. 2014; Jones and Vaupel 2017).

2.8 Immortality of Bacterial, Stem and Cancer Cells Show Similarity in Gene Expression Profiles

Horizontal gene transfer is a key issue in the existence and evolution of prokaryotes. It restores genes inactivated by stochastic mutations, thereby preventing irreversible deterioration of genomes. Although prokaryotes reproduce asexually, the genome sequence analyses show that they frequently undergo genetic recombination, apparently by horizontal gene transfer (Takeuchi et al. 2014).

Tumorigenesis is often considered as reversal of multicellular organism cells towards unicellularity. Sustained proliferation, suppression of cell death, small size and glycolytic energy generation are some of the common hallmarks of tumors and bacteria. Another remarkable finding is that the ancient genes of unicellular origin are highly expressed in tumors. Phylogenetic analyses revealed a close association between the time of gene origin and expression in transformed cells. Genes conserved in prokaryotes were strongly up-regulated, while genes of metazoan origin were inactivated in transformed cells (Trigos et al. 2017). It is noteworthy that total number of bacteria in a man of an average constitution (70 kg) is around 40 trillions (3.8×10^{13})

which is comparable with the number of all human cells (Sender et al. 2016). The immortal nature of bacteria and cancer cells makes them exclusive research objects, especially keeping in mind that the bacteria inhabiting our body are 'friendly and personalized' ones. Moreover, transformed cells become malignant only if they activate mutator genes, i.e., the genes accelerating the mutations of other genes. If only non-mutator genes are modified, the neoplasms could remain benign (Zhu et al. 2022).

2.9 Concluding Remarks

Aging as 'a process inevitably occurring in all cells and organisms' needs to be revised. In the early evolution, immortality was the dominating lifestyle. In these populations, aging emerged later and served for the quick cleaning of damaged constituents accumulated in stressful conditions. There is an obvious similarity between bacterial and immortal cells of multicellular organisms (e.g., stem and cancer cells). Bacteria and some unicellular organisms are immortal in normal living conditions but can purposely age in stressful environments. In contrast, the cells of multicellular organisms age in normal conditions and become immortal only after stressful struggle against the differentiation factors. Definitely, there are enigmatic genetic links between aging and immortality coming from the early times of life origin. The mechanistic damage dilution in cell divisions seems relevant to explain aging and immortality in prokaryotes and metazoans. Multicellular organisms usually sustain the high survival around puberty, i.e., by the decrease in organismal cell division. Moreover, almost all long-lived species (hydra, ocean hog, whales, Greenland sharks, turtles, naked mole-rats, etc.) are growing through the whole life.

Ethics All the materials and data described in the book have been approved by Ethical Committee for Animal Experimentation or Helsinki Committee for Research on Humans of the organizations, where corresponding investigations have been done.

References

Ackermann M, Chao L, Bergstrom CT, Doebeli M (2007) On the evolutionary origin of aging. Aging Cell 6(2):235–244. https://doi.org/10.1111/j.1474-9726.2007.00281.x. PMID: 17376147

Bell EA, Boehnke P, Harrison TM, Mao WL (2015) Potentially biogenic carbon preserved in a 4.1 billion-year-old zircon. Proc Natl Acad Sci USA 112(47):14518–14521. https://doi.org/10.1073/pnas.1517557112. PMID: 26483481

Betts HC, Puttick MN, Clark JW, Williams TA, Donoghue PCJ, Pisani D (2018) Integrated genomic and fossil evidence illuminates life's early evolution and eukaryote origin. Nat Ecol Evol. 2(10):1556–1562. https://doi.org/10.1038/s41559-018-0644-x. PMID: 30127539

Breaker RR, Joyce GF (2014) The expanding view of RNA and DNA function. Chem Biol 21(9):1059–1065. https://doi.org/10.1016/j.chembiol.2014.07.008. PMID: 25237854

Budovsky A, Fraifeld VE, Aronov S (2011) Linking cell polarity, aging and rejuvenation. Biogerontology 12(2):167–175. https://doi.org/10.1007/s10522-010-9305-4. PMID: 20978937

Coelho M, Dereli A, Haese A, Kühn S, Malinovska L, DeSantis ME, Shorter J, Alberti S, Gross T, Tolić-Nørrelykke IM (2013) Fission yeast does not age under favorable conditions, but does so after stress. Curr Biol 23(19):1844–1852. https://doi.org/10.1016/j.cub.2013.07.084. PMID: 24035542

Coelho M, Lade SJ, Alberti S, Gross T, Tolić IM (2014) Fusion of protein aggregates facilitates asymmetric damage segregation. PLoS Biol 12(6):e1001886. https://doi.org/10.1371/journal.pbio.1001886. PMID: 24936793

Damer B, Deamer D (2020) The hot spring hypothesis for an origin of life. Astrobiology 20(4):429–452. https://doi.org/10.1089/ast.2019.2045. PMID: 31841362

Eswara PJ, Ramamurthi KS (2017) Bacterial cell division: nonmodels poised to take the spotlight. Annu Rev Microbiol 71:393–411. https://doi.org/10.1146/annurev-micro-102215-095657. PMID: 28697666

Hirsch HR (1978) The waste-product theory of aging: waste dilution by cell division. Mech Ageing Dev 8(1):51–62. https://doi.org/10.1016/0047-6374(78)90006-4. PMID: 692175

Johnston WK, Unrau PJ, Lawrence MS, Glasner ME, Bartel DP (2001) RNA-catalyzed RNA polymerization: accurate and general RNA-templated primer extension. Science 292(5520):1319–1325. https://doi.org/10.1126/science.1060786. PMID: 11358999

Jones OR, Scheuerlein A, Salguero-Gómez R, Camarda CG, Schaible R, Casper BB, Dahlgren JP, Ehrlén J, García MB, Menges ES, Quintana-Ascencio PF, Caswell H, Baudisch A, Vaupel JW (2014) Diversity of ageing across the tree of life. Nature 505(7482):169–173. https://doi.org/10.1038/nature12789. PMID: 24317695

Jones OR, Vaupel JW (2017) Senescence is not inevitable. Biogerontology 18(6):965–971. https://doi.org/10.1007/s10522-017-9727-3. PMID: 28849291

Jong LW, Fujiwara T, Hirooka S, Miyagishima SY (2021) Cell size for commitment to cell division and number of successive cell divisions in cyanidialean red algae. Protoplasma 258(5):1103–1118. https://doi.org/10.1007/s00709-021-01628-y. PMID: 33675395

Joyce GF (2009) Evolution in an RNA world. Cold Spring Harb Symp Quant Biol 74:17–23. https://doi.org/10.1101/sqb.2009.74.004. PMID: 19667013

Joyce GF, Szostak JW (2018) Protocells and RNA self-replication. Cold Spring Harb Perspect Biol 10(9):a034801. https://doi.org/10.1101/cshperspect.a034801. PMID: 30181195

Kennedy BK, Austriaco NR Jr, Guarente L (1994) Daughter cells of Saccharomyces cerevisiae from old mothers display a reduced life span. J Cell Biol 127(6 Pt 2):1985–1993. https://doi.org/10.1083/jcb.127.6.1985. PMID: 7806576

Kunnev D (2020) Origin of life: the point of no return. Life (Basel) 10(11):269. https://doi.org/10.3390/life10110269. PMID: 33153087

Larocca D, Lee J, West MD, Labat I, Sternberg H (2021) No time to age: uncoupling aging from chronological time. Genes (Basel) 12(5):611. https://doi.org/10.3390/genes12050611. PMID: 33919082

Lilley DM, Sutherland J (2011) The chemical origins of life and its early evolution: an introduction. Philos Trans R Soc Lond B Biol Sci 366(1580):2853–2856. https://doi.org/10.1098/rstb.2011.0133. PMID: 21930575

Lincoln TA, Joyce GF (2009) Self-sustained replication of an RNA enzyme. Science 323(5918):1229–1232. https://doi.org/10.1126/science.1167856. PMID: 19131595

Munné-Bosch S (2015) Senescence: is it universal or not? Trends Plant Sci 20(11):713–720. https://doi.org/10.1016/j.tplants.2015.07.009. PMID: 26442681

Nakaoka H, Wakamoto Y (2017) Aging, mortality, and the fast growth trade-off of Schizosaccharomyces pombe. PLoS Biol 15(6):e2001109. https://doi.org/10.1371/journal.pbio.2001109. PMID: 28632741

Nudler E, Mironov AS (2004) The riboswitch control of bacterial metabolism. Trends Biochem Sci 29(1):11–17. https://doi.org/10.1016/j.tibs.2003.11.004. PMID: 14729327

Pen I, Flatt T (2021) Asymmetry, division of labour and the evolution of ageing in multicellular organisms. Philos Trans R Soc Lond B Biol Sci 376(1823):20190729. https://doi.org/10.1098/rstb.2019.0729. PMID: 33678014

Preiner M, Asche S, Becker S, Betts HC, Boniface A, Camprubi E, Chandru K, Erastova V, Garg SG, Khawaja N, Kostyrka G, Machné R, Moggioli G, Muchowska KB, Neukirchen S, Peter B, Pichlhöfer E, Radványi Á, Rossetto D, Salditt A, Schmelling NM, Sousa FL, Tria FDK, Vörös D, Xavier JC (2020) The future of origin of life research: bridging decades-old divisions. Life (Basel) 10(3):20. https://doi.org/10.3390/life10030020. PMID: 32110893

Proenca AM, Rang CU, Buetz C, Shi C, Chao L (2018) Age structure landscapes emerge from the equilibrium between aging and rejuvenation in bacterial populations. Nat Commun 9(1):3722. https://doi.org/10.1038/s41467-018-06154-9. PMID: 30213942

Proenca AM, Rang CU, Qiu A, Shi C, Chao L (2019) Cell aging preserves cellular immortality in the presence of lethal levels of damage. PLoS Biol 17(5):e3000266. https://doi.org/10.1371/journal.pbio.3000266. PMID: 31120870

Rang CU, Proenca A, Buetz C, Shi C, Chao L (2018) Minicells as a damage disposal mechanism in Escherichia coli. mSphere 3(5):e00428-18. https://doi.org/10.1128/mSphere.00428-18. PMID: 30232168

Saladino R, Crestini C, Pino S, Costanzo G, Di Mauro E (2012) Formamide and the origin of life. Phys Life Rev 9(1):84–104. https://doi.org/10.1016/j.plrev.2011.12.002. PMID: 22196896

Sender R, Fuchs S, Milo R (2016) Revised estimates for the number of human and bacteria cells in the body. PLoS Biol 14(8):e1002533. https://doi.org/10.1371/journal.pbio.1002533. PMID: 27541692

Sheldrake AR (2022) Cellular senescence, rejuvenation and potential immortality. Proc Biol Sci 289(1970):20212434. https://doi.org/10.1098/rspb.2021.2434. PMID: 35232226

Steiner UK (2021) Senescence in bacteria and its underlying mechanisms. Front Cell Dev Biol 9:668915. https://doi.org/10.3389/fcell.2021.668915. PMID: 34222238

Stewart EJ, Madden R, Paul G, Taddei F (2005) Aging and death in an organism that reproduces by morphologically symmetric division. PLoS Biol 3(2):e45. https://doi.org/10.1371/journal.pbio.0030045. PMID: 15685293

Takeuchi N, Kaneko K, Koonin EV (2014) Horizontal gene transfer can rescue prokaryotes from Muller's ratchet: benefit of DNA from dead cells and population subdivision. G3 (Bethesda) 4(2), 325–339. https://doi.org/10.1534/g3.113.009845. PMID: 24347631

Taranukha D, Budovsky A, Gobshtis N, Braiman A, Porat Z, Aronov S, Fraifeld VE (2012) Co-regulation of polar mRNA transport and lifespan in budding yeast Saccharomyces cerevisiae. Cell Cycle 11(22):4275–4280. https://doi.org/10.4161/cc.22659. PMID: 23111244

Trigos AS, Pearson RB, Papenfuss AT, Goode DL (2017) Altered interactions between unicellular and multicellular genes drive hallmarks of transformation in a diverse range of solid tumors. Proc Natl Acad Sci USA 114(24):6406–6411. https://doi.org/10.1073/pnas.1617743114. PMID: 28484005

Wang P, Robert L, Pelletier J, Dang WL, Taddei F, Wright A, Jun S (2010) Robust growth of Escherichia coli. Curr Biol 20(12):1099–1103. https://doi.org/10.1016/j.cub.2010.04.045. PMID: 20537537

Weissman A (2021) Essays upon heredity and kindred biological problems, 2nd ed. Clarendon Press; Oxford, UK: 1891 (cited in Larocca et al. 2021)

Winkler J, Seybert A, König L, Pruggnaller S, Haselmann U, Sourjik V, Weiss M, Frangakis AS, Mogk A, Bukau B (2010) Quantitative and spatio-temporal features of protein aggregation in *Escherichia coli* and consequences on protein quality control and cellular ageing. EMBO J 29(5):910–923. https://doi.org/10.1038/emboj.2009.412. PMID: 20094032

Zhu S, Wang J, Zellmer L, Xu N, Liu M, Hu Y, Ma H, Deng F, Yang W, Liao DJ (2022) Mutation or not, what directly establishes a neoplastic state, namely cellular immortality and autonomy, still remains unknown and should be prioritized in our research. J Cancer 13(9):2810–2843. https://doi.org/10.7150/jca.72628. PMID: 35912015

Chapter 3
Cellular Learning, Decision-Making, and Rejuvenation

Abstract Fundamental principles of information processing, learning, and decision-making are shared among diverse species of prokaryotes and eukaryotes. Current definitions of consciousness and intelligence are overly anthropocentric. Yet, consciousness and intelligence could be ubiquitous and attributable to all unicellulars and cells of multicellular organisms. The first primitive intelligence structures apparently originated in prokaryotes and have likely undergone further development upon transition to unicellular eukaryotes. Most probably that memory, learning, and decision-making capabilities, as essential and genetically fixed traits, have also been retained in the cells of multicellular organisms. A cell can store experience-gained information and use it in a goal-oriented manner. The ability to remember, learn and make logically correct prognosis and decisions pinpoints the existence of cellular intelligence. Although explicit whereabouts and mode of action remain unknown, it is not excluded that there are 'molecular brain(s)' and subordinate systems. In addition to its well-established functions, DNA offers the potential of being that molecular computer, a structure enabling cellular decision-making. Moreover, the existence of biological life for billions of years leaves little doubt that a cell has all necessary resources for rejuvenation. There should be a molecular Rejuvenation Center and Fountain of Youthfulness in every cell, alas, both closed most of the time and in most of the cells. However, induced pluripotency is an excellent example of showing that the cellular institutions of rejuvenation could be activated by external signals. It also supports the paradigm that a cell has an inherent rejuvenation program with a thermodynamically advantageous landscapes of realization, allowing a differentiated cell to find the way back to pluripotency in confusing environments. Whether these structures could be identified and used to learn a cell to be ever-young and share this trait with others remains to be elucidated.

Keywords Cellular computing · Decision-making · DNA methylation · Epigenetic aging clock · Cellular pluripotency · Cellular rejuvenation

K. K. Muradian and V. E. Fraifeld, *Rejuvenation and Longevity*, Healthy Ageing and Longevity 20, https://doi.org/10.1007/978-3-031-64995-0_3

Clever lives longer.

—Hans Friedenthal 1910

3.1 Cognitive Abilities in the World of Cells

Individuals and species that possess more cleverness have obvious advantages in the struggle for existence and thus would prosper. Consequently, correct and quick decision-making and other intelligence-associated features have apparently been a crucial issue since the times of protocells and prokaryotes. Indeed, among many other amazing tricks, bacteria exhibit cognitive abilities, constantly monitor both external and internal environments, compute input/output signals and organize extensive multicellular communications, e.g. so called 'quorum sensing' (Zeng et al. 2023). Bacteria are familiar with altruistic technics of collective defense and attack, apply sophisticated mechanisms of collaboration and often enforce multicellular organisms to play by the bacterial rules (Shapiro 2007). However, many scientists are still quite skeptical about cognitive abilities of unicellulars as well as of the cells of multicellular organisms, and long-lasting debates regarding cell learning, decision-making, and intelligence continue to generate profound interest in the field (Dexter et al. 2019; Brette 2021). Needless to say that much less is known about the aging and rejuvenation aspects of this issue.

Behavior changes that occur as a result of experience, commonly referred to as learning, are lifetime acquired adaptive traits. Unlike genetic selection, which operates over multiple generations, learning mitigates responding to environmental challenges on much faster time scales (Baluška and Levin 2016; Gershman et al. 2021). Empirical research suggests that there are no principal qualitative differences in the complexity of behavior between aneural and neural organisms. Cognitive behavior can be observed across the entire phylogenetic tree, including prokaryotes, ciliates, and fungi (van Duijn 2017). While still subject to debate, memory, learning, and decision-making appear to be inherent and ubiquitous cellular attributes common for most, if not all, organisms (Trewavas and Baluška 2011; Reid et al. 2015; Dussutour 2021; Gershman et al. 2021).

Testing amoeboid slime mold, *Physarum polycephalum* (*P. polycephalum*), by the maze of 'the shortest path in the Towers of Hanoi' revealed its ability to construct minimal length paths through the initial maze as well as after new passages are introduced (Reid and Beekman 2013). Even more impressive are the results of experiments with the well-known test of 'multi-armed bandit' adapted from human studies. In this test, the examinee should make quick decisions regarding which of the several slot-machines to invest in, aiming to maximize the rewards. The rewards may either be frequent but small or infrequent but large. The slime mold demonstrated the remarkable ability to make correct decisions by comparing multiple options and considering information on both reward frequency and magnitude (Reid et al. 2016). To observe an amoeboid successfully coping with the task which challenges many humans, could be amusing up to absurdity. Yet, it also instills a sense of pride for our primordial ancestors.

The world of unicellulars exhibits amazing resourcefulness at minimal structural support. Amoeba, for instance, lacking eyes and ears, 'sees' and 'hears' perfectly. It can effortlessly differentiate between predators and prey, responding by swiftly evading the former and attacking the latter. Bacteria, amoebas and many other unicellular organisms, when faced with danger, generate stress response signals that can be sensed, analyzed, and remembered by their mates. Through brief communication, *P. polycephalum*, for example, is able to receive ample warning instructions from a stressed clone to actively avoid previously explored toxic environments (Briard et al. 2020). Studies on memory duration revealed that a slime molds are capable of retaining habituation information even after one month of dormancy (Boussard et al. 2019). Moreover, amoeba widely utilizes tricks that are not commonly used by humans, including the development of mating strategies, their regular assessment and correction of possible mistakes (Clark 2013). Unicellular eukaryotes and prokaryotes displayed remarkable wit and resourcefulness in many other analogous experiments (Baluška and Levin 2016; Boisseau et al. 2016; Vogel and Dussutour 2016; Reid et al. 2016; De la Fuente et al. 2019; Brette 2021; Gershman et al. 2021). As noted by Shapiro, a prokaryote genetics expert with forty years of experience, "*bacteria are small but not stupid*" (Shapiro 2007). All in all, the available data strongly suggests that the principles of information processing and decision-making are conserved throughout evolution. New findings will further reshape the old concepts about different types of minds and intelligence, their composition and their origin.

3.2 DNA and Proteins in Cellular Computing

The selective pressure on optimizing decision-making processes should act as a driving force for the broader utilization of the past experience (memory) and efficient methods of information processing. Considering the complexity of multidimensional external and intracellular environments, such extensive information processing necessitates large-scale computation. In pursuit of the putative cellular computer, it is challenging to overlook the role of DNA: this macromolecule is widely recognized for its ability to function as billions of interconnected nanoprocessors (de Silva and Uchiyama 2007; Ji 2022).

DNA serves several crucial functions in living organisms. Firstly, it acts as a depositary or library for storing genetic information. Secondly, it functions as the machinery for transcription, allowing to produce RNA copies. Thirdly, DNA shares its genetic information with sister cells through replication. Lastly, DNA transfers genetic information across generations via zygotes. In addition to these well-established functions, *there are two other hypothetical functions that DNA may play: computation and participation in cellular decision-making*. The study of molecular computation in living cells, though still in its infancy, holds great potential for monitoring biological processes and predicting cellular behavior (Li et al. 2017). DNA molecules can be manipulated to perform mathematical computations even in a test tube. Remarkably, the computational capacity of DNA has been found to be orders of magnitude greater than that of the most powerful human-built silicon computers (Benenson et al. 2003; Fan et al. 2020).

However, it is not only DNA that possesses computational capabilities within a cell. It looks quite plausible that many proteins in living cells may primarily function in the transfer and processing of information rather than just chemical substrate transformation. For example, the neuron-like r-protein networks in ribosomes are capable of performing a variety of computational tasks, from amplification and integration to information storage (Bray 1995; Timsit and Grégoire 2021). This highlights the intriguing potential for information processing within cellular systems beyond DNA alone.

3.3 Cellular Command-Analytical Hub

Molecular carriers of memory, learning, and decision-making are expected to act in a cooperative manner. If so, when assembled together, they can form the framework of a *command-analytical hub* with a wide spectrum of functionalities. We hypothesize that key structures of the command-analytical hubs could be located within the well-protected nuclear/nucleolar heterochromatin or in remote plasmids, while mobile units could be associated with small RNAs, transposons, introns, and other similar elements. Then, a great amount of so-called 'junk' DNA and introns in the genome, with seemingly no purpose in gene expression, could be indispensable in cellular computation. Notably, transposable elements (TEs) occupy a considerable portion of the mammalian genome (see Chap. 7) and their abundancy displays a significant positive correlation with mammalian longevity (data not shown; paper in preparation). To be honest, the latter finding was a bit of a surprise for us. Yet, given some proportionality between the TE content and the quality of cellular computation, our observation no longer looks so strange. Whether DNA is truly involved in large-scale computing, predicting, and supervising all, or nearly all, cellular events, remains disputable and represents a point for future thorough investigation.

3.4 DNA Methylation and Epigenetic Aging Clock

The concept of 'all knowing and all mighty' command-analytical hub could be relevant for explanation of many situations otherwise difficult to comprehend. For instance, the dilemma of using DNA methylation as an epigenetic clock. Methylome-based aging clocks are successfully used to predict biological age and assess efficiency of anti-aging and rejuvenative interventions (Horvath 2013) (see Chap. 11,

Table 11.1). However, are changes in DNA methylation an independent process or are they a consequence of related cellular events? Methylome could hardly be supplied by signals pervasive enough for monitoring aging rate. But even if there could be channels of information transfer, how the avalanche of age-associated changes would be processed by a relatively simple methylation apparatus? Participation of the command-analytical hub as the third party could significantly simplify the situation. Indeed, being the best informed and authoritative cellular structure, the command-analytical hub could indicate which CpG sites should be methylated or demethylated (see Chap. 11).

3.5 Cellular Pluripotency, Decision-Making, and Rejuvenation

The discovery of induced pluripotency (Takahashi and Yamanaka 2006) has been perceived by many as a long-awaited scientific miracle in anti-aging and rejuvenation. Indeed, who could imagine that it would be possible to not only slow down or suspend but even turn back the hurricane of age-related changes accumulated by adult cells? Incessant changes that occur all around—in the metabolome, subcellular compartments, in the mitochondrial and nuclear genomes, everywhere! Who would think that activation of few transcription factors or introducing the defined cocktails of small molecules could be enough for initiating a cell's journey back to pluripotency and rejuvenation (Takahashi and Yamanaka 2006; Hou et al. 2013; Knyazer et al. 2021; Guan et al. 2022)? *As such, induced pluripotency (iP) is a remarkable example of cellular intelligence, resourcefulness, and astute decision-making*. Unfortunately, because of myriad of still unresolved technical issues, the anticipated cascade of iP-based groundbreaking discoveries and their subsequent application in medical and biological practices remains thus far as an almost unrealized opportunity.

3.6 How a Cell Finds the Way Back to Pluripotency and Rejuvenation?

It is just a mystery how an aged cell finds the way back to complete rejuvenation through iP. How can a cell return to the starting point on its own, after passing the incredibly complex series of transformations and intricate twists and turns along its life path?! Is the trajectory of countless changes experienced by an aging cell recorded in detail in the history of its life? Could a cell, as a precautionary measure, mark the path back to pluripotency? Perhaps the simplest explanation for this enigmatic phenomenon could be linked to the Waddingtonian thermodynamic landscapes (Li and Wang 2013; Sáez et al. 2022). Replicative rejuvenation most likely emerged in early protocells and occupied thermodynamically advantageous landscapes (see

Chap. 11, the second law of rejuvenation). At those times, there should be plenty of opportunities for choosing the most favorable thermodymanic routes that enable partially damaged cells to return spontaneously to the starting points, and thus to repair and rejuvenate. These landscapes have obviously been vitally important to be properly conserved in the genome and inherited through generations. Hence, for a cell, it is sufficient to locate the relevant thermodynamic valley and 'roll down' by the minimal free energy trajectory. It is remarkable that the cells undergoing reprogramming can return not only to the points of their origin (usually partially differentiated stem cells), but also overcome hurdles in the final stages of reprogramming way, formerly unknown to them. This *terra incognito* ultimately brings all types of the cells to the same destination point on their journey back—a state of complete dedifferentiation and pluripotency.

3.7 Cellular Rejuvenation Center and Fountain of Youthfulness

The hypothetical command-analytical hub should be thoroughly scrutinized. Indeed, we currently do not know explicitly where it is located and how it operates. However, we know almost for sure that a cell has inherent machinery for memory, learning, and goal-oriented decision-making. The carriers of these traits could hardly be completely isolated, they have to cooperate. Assembled in the frame of a command-analytical hub, they could become the most powerful regulatory-executive center in a cell. These structures have possibly been established at the dawn of life origin and could be conservative enough to keep cells focused on pluripotency and rejuvenation. Moreover, the very existence of biological life during billions of years leaves little doubt that a cell has all necessary resources for complete rejuvenation. In the course of evolution cells underwent an incredible variety of modifications. However, all cells that survived should retain the ability of complete rejuvenation, otherwise they will disappear in the mist of evolution. Therefore, the very continuity of biological life proves that there should be inherent molecular Rejuvenation Center(s) and corresponding Fountain(s) of Youthfulness in every normal cell. In addition, induced pluripotency proves that these cellular institutions could be activated by external signals. Alas, these institutions are 'bureaucratic' enough to be closed most of the time and in most of the cells. Thus, it is of utmost significance to find a way of how to communicate and trade-off with these yet hypothetical structures.

3.8 Concluding Remarks

We are used to thinking that cells are merely basic and relatively simple building blocks, and only the huge number and diversity of these cells ensure the existence of conscious organisms. However, it is an outdated delusion. A cell is far from being that simple. Indeed, every cell is a wholesome living entity, with a complete set of inherent traits for intelligence and rejuvenation. The concepts of a 'molecular brain' or 'cellular cognition and consciousness' are gaining increasing recognition and acceptance. There is a quickly growing consensus that a cell can learn and make logically correct predictions and decisions. It exhibits a goal-oriented behavior, with rejuvenation and longevity in the list of priorities. That is, *a cell has a goal—to keep the biological life going on!*

The idea of 'molecular brain' raises the intriguing question of whether it is possible to rank differentiated cells or organs in terms of their 'cleverness', like we compare different animal species or individuals. How are cellular, tissue and organismal intelligences related? These inquiries become even more complex in the context of pathologies or aging and rejuvenation. For example, can we say that a cancer cell 'has lost its mind' through damaged command-analytical hub and false hopes to escape from age-associated chaos and imminent collective death by replicative rejuvenation and multiplication of same-minded mates? Whatsoever, the links between cellular, tissue and organismal wits, full of exiting discoveries, remain open for exploration and discussion.

The lack of knowledge regarding the explicit nature and subcellular localization of the hypothetical 'molecular brain' and related structures poses a significant challenge to their study and potential application. However, this does not negate their suggested role in various processes, aging and rejuvenation included. Cellular memory, learning and decision-making ensure a certain degree of intellectual freedom for both unicellulars and cells within multicellular organisms. However, whether these structures are the master keepers of the keys to unlocking the cellular 'fountain of youthfulness' remains unexplored.

Consciousness and goal-oriented behavior allow us to look at biological objects from different perspectives. When conducting experiments with tissues and cells, it is important to consider them as conscious entities with their own memories and life histories. Do they experience pain when mistreated? Can they forgive insults or are they simply ashamed of the insulter? No answers from that shore yet.

Overenthusiastic visons and assumptions are worthwhile if they also indicate ways of empirical checking and translation of the obtained results into realistic deals. The discussed concepts are largely 'dare' and highly disputable hypotheses. A lot remains to be done, and only further research will show their validity and applicability.

Ethics All the materials and data described in the book have been approved by Ethical Committee for Animal Experimentation or Helsinki Committee for Research on Humans of the organizations, where corresponding investigations have been done.

References

Baluška F, Levin M (2016) On having no head: cognition throughout biological systems. Front Psychol 7:902. https://doi.org/10.3389/fpsyg.2016.00902. PMID: 27445884

Benenson Y, Adar R, Paz-Elizur T, Livneh Z, Shapiro E (2003) DNA molecule provides a computing machine with both data and fuel. Proc Natl Acad Sci USA 100(5):2191–2196. https://doi.org/10.1073/pnas.0535624100. PMID: 12601148

Boisseau RP, Vogel D, Dussutour A (2016) Habituation in non-neural organisms: evidence from slime moulds. Proc Biol Sci 283(1829):20160446. https://doi.org/10.1098/rspb.2016.0446. PMID: 27122563

Boussard A, Delescluse J, Pérez-Escudero A, Dussutour A (2019) Memory inception and preservation in slime moulds: the quest for a common mechanism. Philos Trans R Soc Lond B Biol Sci 374(1774):20180368. https://doi.org/10.1098/rstb.2018.0368. PMID: 31006372

Bray D (1995) Protein molecules as computational elements in living cells. Nature 376(6538):307–312. https://doi.org/10.1038/376307a0.Erratum.In:Nature1995;378(6555):419. PMID: 7630396

Brette R (2021) Integrative neuroscience of paramecium, a "swimming neuron". eNeuro 8(3). https://doi.org/10.1523/ENEURO.0018-21.2021. PMID: 33952615

Briard L, Goujarde C, Bousquet C, Dussutour A (1802) Stress signalling in acellular slime moulds and its detection by conspecifics. Philos Trans R Soc Lond B Biol Sci 375(1802):20190470. https://doi.org/10.1098/rstb.2019.0470. PMID: 32420856

Clark KB (2013) Ciliates learn to diagnose and correct classical error syndromes in mating strategies. Front Microbiol 4:229. https://doi.org/10.3389/fmicb.2013.002294:229.doi:10.3389/fmicb.2013.00229. PMID: 23966987

De la Fuente IM, Bringas C, Malaina I, Fedetz M, Carrasco-Pujante J, Morales M, Knafo S, Martínez L, Pérez-Samartín A, López JI, Pérez-Yarza G, Boyano MD (2019) Evidence of conditioned behavior in amoebae. Nat Commun 10(1):3690. https://doi.org/10.1038/s41467-019-11677-w. PMID: 31417086

de Silva AP, Uchiyama S (2007) Molecular logic and computing. Nat Nanotechnol 2(7):399–410. https://doi.org/10.1038/nnano.2007.188. PMID: 18654323

Dexter JP, Prabakaran S, Gunawardena J (2019) A complex hierarchy of avoidance behaviors in a single-cell eukaryote. Curr Biol 29(24):4323-4329.e2. https://doi.org/10.1016/j.cub.2019.10.059.Erratum.In:CurrBiol.2020;30(11):2205. PMID: 31813604

Dussutour A (2021) Learning in single cell organisms. Biochem Biophys Res Commun 564:92–102. https://doi.org/10.1016/j.bbrc.2021.02.018. PMID: 33632547

Fan D, Wang J, Wang E, Dong S (2020) Propelling DNA computing with materials' power: recent advancements in innovative DNA logic computing systems and smart bio-applications. Adv Sci (Weinh) 7(24):2001766. https://doi.org/10.1002/advs.202001766. PMID: 33344121

Gershman SJ, Balbi PE, Gallistel CR, Gunawardena J (2021) Reconsidering the evidence for learning in single cells. Elife 10:e61907. https://doi.org/10.7554/eLife.61907. PMID: 33395388

Guan J, Wang G, Wang J, Zhang Z, Fu Y, Cheng L, Meng G, Lyu Y, Zhu J, Li Y, Wang Y, Liuyang S, Liu B, Yang Z, He H, Zhong X, Chen Q, Zhang X, Sun S, Lai W, Shi Y, Liu L, Wang L, Li C, Lu S, Deng H (2022) Chemical reprogramming of human somatic cells to pluripotent stem cells. Nature 605(7909):325–331. https://doi.org/10.1038/s41586-022-04593-5. PMID: 35418683

Horvath S (2013) DNA methylation age of human tissues and cell types. Genome Biol 14(10):R115. https://doi.org/10.1186/gb-2013-14-10-r115.Erratum.In:GenomeBiol.2015;16:96. PMID: 24138928

Hou P, Li Y, Zhang X, Liu C, Guan J, Li H, Zhao T, Ye J, Yang W, Liu K, Ge J, Xu J, Zhang Q, Zhao Y, Deng H (2013) Pluripotent stem cells induced from mouse somatic cells by small-molecule compounds. Science 341(6146):651–654. https://doi.org/10.1126/science.1239278. PMID: 23868920

Ji S (2022) Molecular mechanisms of encoding and decoding information in cell computing. Biosystems 219:104715. https://doi.org/10.1016/j.biosystems.2022.104715. PMID: 35690290

Knyazer A, Bunu G, Toren D, Mracica TB, Segev Y, Wolfson M, Muradian KK, Tacutu R, Fraifeld VE (2021) Small molecules for cell reprogramming: a systems biology analysis. Aging (Albany NY) 13(24):25739–25762. https://doi.org/10.18632/aging.203791. PMID: 34919532

Li C, Wang J (2013) Quantifying Waddington landscapes and paths of non-adiabatic cell fate decisions for differentiation, reprogramming and transdifferentiation. J R Soc Interface 10(89):20130787. https://doi.org/10.1098/rsif.2013.0787. PMID: 24132204

Li J, Green AA, Yan H, Fan C (2017) Engineering nucleic acid structures for programmable molecular circuitry and intracellular biocomputation. Nat Chem 9:1056–1067. https://doi.org/10.1038/nchem.2852. PMID: 29064489

Reid CR, Beekman M (2013) Solving the towers of Hanoi—how an amoeboid organism efficiently constructs transport networks. J Exp Biol 216(Pt 9):1546–1551. https://doi.org/10.1242/jeb.081158. PMID: 23307798

Reid CR, Garnier S, Beekman M, Latty T (2015) Information integration and multiattribute decision making in non-neuronal organisms. Anim Behav 100:44–50. https://doi.org/10.1016/j.anbehav.2014.11.010

Reid CR, MacDonald H, Mann RP, Marshall JA, Latty T, Garnier S (2016) Decision-making without a brain: how an amoeboid organism solves the two-armed bandit. J R Soc Interface 13(119):20160030. https://doi.org/10.1098/rsif.2016.0030. PMID: 27278359

Sáez M, Briscoe J, Rand DA (2022) Dynamical landscapes of cell fate decisions. Interface Focus 12(4):20220002. https://doi.org/10.1098/rsfs.2022.0002. PMID: 35860004

Shapiro JA (2007) Bacteria are small but not stupid: cognition, natural genetic engineering and socio-bacteriology. Stud Hist Philos Biol Biomed Sci 38(4):807–819. https://doi.org/10.1016/j.shpsc.2007.09.010. PMID: 18053935

Takahashi K, Yamanaka S (2006) Induction of pluripotent stem cells from mouse embryonic and adult fibroblast cultures by defined factors. Cell 126(4):663–676. https://doi.org/10.1016/j.cell.2006.07.024. PMID: 16904174

Timsit Y, Grégoire SP (2021) Towards the idea of molecular brains. Int J Mol Sci 22(21):11868. https://doi.org/10.3390/ijms222111868. PMID: 34769300

Trewavas AJ, Baluška F (2011) The ubiquity of consciousness. EMBO Rep 12(12):1221–1225. https://doi.org/10.1038/embor.2011.218. PMID: 22094270

van Duijn M (2017) Phylogenetic origins of biological cognition: convergent patterns in the early evolution of learning. Interface Focus 7(3):20160158. https://doi.org/10.1098/rsfs.2016.0158. Epub 2017 Apr 21. PMID: 28479986; PMCID: PMC5413897

Vogel D, Dussutour A (2016) Direct transfer of learned behaviour via cell fusion in non-neural organisms. Proc Biol Sci 283(1845):20162382. https://doi.org/10.1098/rspb.2016.2382. PMID: 28003457

Zeng X, Zou Y, Zheng J, Qiu S, Liu L, Wei C (2023) Quorum sensing-mediated microbial interactions: mechanisms, applications, challenges and perspectives. Microbiol Res 273:127414. https://doi.org/10.1016/j.micres.2023.127414. PMID: 37236065

Chapter 4
Sporulation, Meiosis, and Rejuvenation

Abstract Spores are extraordinary resistant life forms. They can exist in contaminated environments for centuries. To assume the spore's longevity potential, it is enough to mention that the longest-existing bacterial spores reach 250 million years. A rejuvenated spore can miraculously return to active life within few minutes of germination. Spores can survive in space journeys, and they are among the potential pioneers who habituated Earth and other planets. At the same time, spores are the simplest and archaistic mode of meiotic rejuvenation. *Bacillus subtilis* (*B. subtilis*) and yeast species (e.g., *Schizosaccharomyces pombe*) often serve as model systems for exploring the spores and sporulation. Remarkably, new progenies are equally rejuvenated despite the differences in parental ages and duration of dormancy. Species from all three domains (Eubacteria, Archaea, and Eukarya) clearly demonstrate that meiotic rejuvenation is an efficient and ubiquitous event in nature, which contributes a lot to the very continuity of life.

Keywords Sporulation · Gametogenesis · Meiosis · Meiotic rejuvenation · Spore dormancy · Spore germination · Fermi paradox · Panspermia

A model should be as simple as it can be but no simpler.

—Albert Einstein

4.1 Sporulation and Spores

Bacterial endospores were independently discovered around 150 years ago by Robert Koch and Louis Pasteur (Mock and Fouet 2001; Gould 2006). Nowadays, the results of electron cryotomography clearly indicate that sporulation-like events have an evolutionary long history, so that the ancient spores could be the last bacterial common ancestor (Tocheva et al. 2016). Moreover, numerous species from all three

K. K. Muradian and V. E. Fraifeld, *Rejuvenation and Longevity*, Healthy Ageing and Longevity 20, https://doi.org/10.1007/978-3-031-64995-0_4

domains—Eubacteria, Archaea, and Eukarya (bacteria, fungi, algae, plants, protozoa, etc.) use sporulation to survive the difficult times and give rise to rejuvenated progeny in favorable conditions (Knight and Goddard 2016; Wang and Montero 2016).

Bacillus subtilis (*B. subtilis*) and related species are the most known model systems for exploring the morphological, biochemical, and genetic determinants of spores. Simple models have apparent advantages. Let us start from the beginning—the initiation of sporulation. It is well known that activation of the defense and repair systems usually requires significant energy-consuming resources. To activate sporulation, it is enough only depriving nutrients. These unfavorable conditions initiate asymmetric cell division which eventually results in the formation of tiny participles, the spores. The most part of spore's volume is occupied by nucleic acids and defense membrane. This simple unit of life can stay indefinitely long in dormancy and give rise to four newly rejuvenated organisms in suitable living conditions (Huang and Hull, 2017; Tu et al. 2021).

Bacteria experience nutrient deficiency most of the time. Like spores, they also developed mechanisms for long surviving in a growth-arrested state, but without the spores' dormant structures. Some bacteria can generate ATP from light during growth arrest and maintain a certain metabolic activity (Yin et al. 2019). Around a hundred bacterial genes have been identified as being linked to bacterial longevity at the quiescent conditions. Among them are the genes for DNA repair, tRNA modification, and the fidelity of protein synthesis (Pechter et al. 2017). The growth arrest and associated expression changes in response to nutrient depletion promote survivability of bacteria and extend their life span (Dworkin and Harwood, 2022).

4.2 Spores Can Exist in Dormancy for Years and More

Spores are highly resistance biological entities. They can survive in contaminated environments for years and centuries (Errington 2003; Segev et al. 2012). There are various species and strains of spores which can endure extraordinary doses of UV and γ-radiation, high pressure, desiccation, freezing, thawing, elevated temperatures, and toxic chemicals, including oxidizing and alkylating agents, aldehydes, halogens, acids and bases (Moeller et al. 2011; Setlow 2014a, b; Malleck et al. 2022). Nearly all samples of the Arctic and Antarctic permafrosts, from 5–10 thousand to 2–3 million years old, contain sporulated bacteria and fungi (Zhang et al. 2013; Brouchkov et al. 2017, 2020). Bacterial spores can exist even longer. In *B. sphaericus*, found in a Dominican amber, the age of spores could reach 25–30 million years (Cano et al. 1994). Bacterial age can show astonishing records in the spores isolated from the insects trapped in salt crystal in the Permian Salado Formation (New Mexico). Their age could reach 250 million years! Complete sequencing of the 16S ribosomal DNA from these spores showed that they could be a part of the lineage of *B. marismortui* and *Virgibacillus pantothenticus* (Vreeland et al. 2000).

Remarkably, 250 million years correspond to a cycle of Pangea—a period of the convergence and divergence of the earth's supercontinent Pangea. Since a spore can exists that much long, we take our liberty to suggest 1 (one) Pangea as a time unit for measuring very long-lasting biological phenomena, with hopes of employing the Pangea units for a future human life span (☺).

4.3 Fermi Paradox, Panspermia, and Spores

From the antiquity to the modern civilizations, it has been believed that skies are inhabited by gods, angles, living creatures, etc. Panspermia is a philosophical idea that such living beings can initiate life on Earth and other planets (Mitton 2022). Yet, the paradox of Enrico Fermi or the 'Great silence problem' claims that if life exists somewhere in the space, it would be somehow observed on Earth (Gray 2015).

Due to the extraordinary resistance, bacteria and viruses have often been the subject of astronomical explorations, e.g., in 'cometary panspermia' programs. Cometary panspermia posits that comets harbor viable biological subjects in the form of bacteria or viruses, thereby leading eventually to the origin and evolution of life on Earth (Wickramasinghe et al. 2020). Direct experiments have demonstrated that under favorable circumstances spores can survive a journey within our solar system. Spores of *B. subtilis* were exposed to the space satellite without protection or with protection against UV such as mixed clay, red sandstone, Martian analogue soil or meteorite powder. Two weeks later, almost all the spores kept in the protected conditions remained alive, in contrast to the unprotected spores of a near-zero survival (Horneck et al. 2001).

4.4 Are Dormant Spores Metabolically Active?

There is an old disputable question: is a spore entirely dormant or does have any metabolic activity? In fact, nucleic acids are complex macromolecules which would hardly exist long without any repair and associated energy support. Yet, there is evidence that bacterial dormant spores may exhibit enzymatic activities (Sanchez-Salas et al. 2011), single-strand DNA breaks repair (Durban et al. 1974), and mobility of phosphorous compounds (Leuschner and Lillford 2000). In dormant spores of *B. subtilis* and *B. megaterium*, 95% of rRNAs were degraded during incubation at 50 °C. However, the spores with a depleted rRNA pool were viable and completed the germination and vegetative growth even faster than the untreated spores (Segev et al. 2012; Korza et al. 2016). Whatever the case, it is indeed amazing how a dormant spore finds energy sources for the existence for years?

4.5 Sporulation and Meiosis

Meiotic rejuvenation is the essence of sporulation. In normal feeding conditions, *S. cerevisiae* propagates by mitotic divisions. Nutrient starvation, however, prompts a yeast cell to enter the meiotic cycle and produce rejuvenated cells. Consequently, *S. cerevisiae* is probably the best studied model for exploration of *meiotic rejuvenation* (Unal et al. 2011; King et al. 2019).

It should be noted that in contrast to mitosis, meiosis is not a quick process. Another important point is that the most intensive removal of accumulated damage occurs virtually at the very end of meiosis (King and Ünal 2020). This could be particularly relevant for long-lived species.

Nutrient depletion-induced sporulation may vary across species. In a filamentous ascomycete, belonging to the yeast family *Saccharomycetaceae*, nutrient depletion generates a strong positive signal for sporulation via the cAMP-PKA (protein A kinase) pathway (Wendland 2020). In fission yeast *S. pombe*, induction of sporulation requires the presence of cells of the opposite sex (Ohtsuka et al. 2022).

4.6 The Genome and Gene Expression in Meiotic Rejuvenation

Nuclear periphery and chromosomes therein are the well-known hotspots of the age-associated damages (King and Ünal 2020). These sites, including enlarged nucleoli, extrachromosomal circular rDNA, shortened telomeres, impaired nuclear pore complexes as well as damaged subcellular organelles and defected proteins, etc. are in active recovery during gametogenesis across eukaryotes. This is accompanied by multiplication and redistribution of transposable elements (see Chap. 7). Correction and elimination of the abovementioned defects with recovery to the normal state is the prime target, though not the only one, of meiotic rejuvenation (Sinclair and Guarente 1997; Neiman 2011; Unal et al. 2011; King et al. 2019; King and Ünal 2020).

S. cerevisiae meiosis results in generation of two types of haploid cells expressing mating MATa and MATα loci. These haploid cells can converge and create three types of diploid cells: a/a, α/α, and a/α. The a/a or α/α diploid cells are unable to sporulate and sporulation is restricted only to the diploid a/α cells, in analogy with the bisexual reproduction in multicellular organisms (Deutschbauer et al. 2002). Once committed, sporulating cells activate the gene cascade which involves ca. one third of genes of the yeast entire genome. Around a quarter of *S. cerevisiae* genes (1600 out of 6200) may be involved in activation of the sporulation process. Compared to the activators of sporulation, the number of downregulated genes is far lower (usually less than few hundreds) (Gupta et al. 2015; Maire et al. 2020).

4.7 Age and Meiotic Rejuvenation

Spores (gametes) display the same replicative potential and are born young, regardless of their parents age. This demonstrates explicitly that *gamete formation fully rejuvenates and resets replicative life span* (Unal et al. 2011; Koch-Bojalad et al. 2021). Time-lapse fluorescence microscopy showed that extrachromosomal rDNA circles, dysfunctional protein aggregates and abnormal core subunits of the nuclear pore complexes are physically sequestered away from chromosomes during the late meiosis (anaphase of meiosis II). Moreover, it was found that the sequestrated materials are collected in a special membrane and eliminated by enzymatic systems (Unal et al. 2011; King et al. 2019; Sing et al. 2022).

However, meiotic cells (meiocytes) from old progenitor cells may encounter serious difficulties in accomplishing the meiosis. Examination of the effects of replicative aging revealed that old mother cells have a decreased ability to initiate the meiotic program. Even if the program is completed, the meiocytes often have defects in chromosome segregation (Boselli et al. 2009). Remarkably, interventions and mutations that extend replicative life span often also increase reproductive life span, thus indicating that the fidelity of meiosis could be controlled by the same pathways that govern aging rate of yeast cells (King et al. 2019).

If the same is true for mammals, then it may explain one of the enigmas of human reproduction— long delay in the final stage of oocyte maturation. Indeed, it is well-known that in mammalian oocyte, meiosis starts in a fetus, i.e., before birth. Once meiosis starts, oocytes quickly pass several stages of the meiosis I—leptotene, zygotene, and pachytene—and are arrested at the dictyotene stage. The arrest is terminated only after reaching puberty, i.e., it can be extended for decades in humans and other long-lived species (Lu and Yu 2015; Stringer et al. 2020; He et al. 2021). Why such a long meiotic program is necessary? One of the explanations is that dictyotene is a time-consuming period of sister chromatid recombination and the efficient recovery of oocyte DNA by homologous recombinational repair (Gartner and Engebrecht 2022). Because the most intensive removal of accumulated damage occurs virtually at the very end of meiosis (King et al. 2019), such a long-standing delay in the final stage of meiosis may allow to start fertilization with minimal age-associated damages.

4.8 Spore Germination

In a nutrient-reach environment, the spores can convert the dormancy to an active cellular state through germination—a process in which haploid spores (gametes) resume vegetative growth by re-entering the mitotic cell cycle through the G1 phase (Moir and Cooper 2015). Spore germination is a multi-step and surprisingly rapid process requiring cardinal alteration in morphology, metabolism, cellular content, etc. (Deutschbauer et al. 2002). First, a spore goes through a process of un-coating by losing its unique wall. The gluconeogenesis pathway, which is important in long periods of starvation for a glucose synthesis from non-carbohydrate precursors, is inhibited immediately after germination starts (Ray and Ye 2013). In parallel, the genes related to the tricarboxylic acid cycle and oxidative phosphorylation become re-activated. At the early stages of germination, expression of about 1000 genes (550 induced and 480 repressed out of approximately 6200 yeast genes) is modified in five minutes after nutrient medium recovery (Joseph-Strauss et al. 2007). After germination, a cell starts synthesizing new macromolecules and activates mitotic cell division (Setlow 2014a, b; Moir and Cooper 2015; Christie and Setlow 2020).

4.9 Concluding Remarks

Spore formation is perhaps the most primordial mechanism of gametogenesis. *B. subtilis* and yeast species are obviously the simplest model systems for exploring this process and have contributed a lot to our understanding of meiotic rejuvenation. In contrast to somatic cells with their decreased ability to combat aging, spores (gametes) have developed robust strategies for damage sequestration and degradation. During meiosis the cells undergo profound resetting of all cellular structures, mostly, nuclear periphery and chromosomes, and eventually get rid of the accumulated damage. Remarkably, the meiocytes originated from the old parent cells preserve the same rejuvenative capacity as the meiocytes originated from the young parent cells. Altogether, this ensures to *minimize damage transfer across the generations*. In essence, *meiosis is the first step and natural mechanism of rejuvenation* (see also Chap. 11).

Ethics All the materials and data described in the book have been approved by Ethical Committee for Animal Experimentation or Helsinki Committee for Research on Humans of the organizations, where corresponding investigations have been done.

References

Boselli M, Rock J, Unal E, Levine SS, Amon A (2009) Effects of age on meiosis in budding yeast. Dev Cell 16(6):844–855. https://doi.org/10.1016/j.devcel.2009.05.013. PMID: 19531355

Brouchkov A, Melnikov V, Kalenova L, Fursova O, Pogorelko G, Potapov V, Fursova N, Ignatov S, Brenner E, Bezrukov V, Muradian K (2017) Permafrost bacteria in biotechnology: biomedical applications. In: Margesin R, Schinner F, Marx J-C, Gerday C (eds) Psychrophiles: from biodiversity to biotechnology. Springer, pp 541–554

Brouchkov A, Melniko V, Griva GI, Kashuba E, Kashuba V, Kabilov M, Fursova O, Bezrukov V, Muradian K et al (2020) Microorganisms from permafrost and their possible applications. In: Yadav AN, Rastegari AA, Yadav N (eds) Microbiomes of extreme environments: biodiversity and biotechnological applications, 1st ed, vol 1. CRC Press, pp 252–277

Cano RJ, Borucki MK, Higby-Schweitzer M, Poinar HN, Poinar GO Jr, Pollard KJ (1994) Bacillus DNA in fossil bees: an ancient symbiosis? Appl Environ Microbiol 60(6):2164–2167. https://doi.org/10.1128/aem.60.6.2164-2167.1994. PMID: 8031102

Christie G, Setlow P (2020) Bacillus spore germination: knowns, unknowns and what we need to learn. Cell Signal 74:109729. https://doi.org/10.1016/j.cellsig.2020.109729. PMID: 32721540

Deutschbauer AM, Williams RM, Chu AM, Davis RW (2002) Parallel phenotypic analysis of sporulation and postgermination growth in *Saccharomyces cerevisiae*. Proc Natl Acad Sci USA 99(24):15530–15535. https://doi.org/10.1073/pnas.202604399. PMID: 12432101

Dworkin J, Harwood CS (2022) Metabolic reprogramming and longevity in quiescence. Annu Rev Microbiol 76:91–111. https://doi.org/10.1146/annurev-micro-041320-111014. PMID: 35417196

Durban E, Grecz N, Farkas J (1974) Direct enzymatic repair of deoxyribonucleic acid single-strand breaks in dormant spores. J Bacteriol 118(1):129–138. https://doi.org/10.1128/jb.118.1.129-138.1974. PMID: 4206867

Errington J (2003) Regulation of endospore formation in *Bacillus subtilis*. Nat Rev Microbiol 1(2):117–126. https://doi.org/10.1038/nrmicro750. PMID: 15035041

Gartner A, Engebrecht J (2022) DNA repair, recombination, and damage signaling. Genetics 220(2):iyab178. https://doi.org/10.1093/genetics/iyab178. PMID: 35137093

Gray RH (2015) The fermi paradox is neither Fermi's nor a paradox. Astrobiology 15(3):195–199. https://doi.org/10.1089/ast.2014.1247. Epub 2015 Feb 26 PMID: 25719510

Gould GW (2006) History of science—spores. J Appl Microbiol 101(3):507–513. https://doi.org/10.1111/j.1365-2672.2006.02888.x. PMID: 16907801

Gupta R, Sadhale PP, Vijayraghavan U (2015) SUB1 plays a negative role during starvation induced sporulation program in *Saccharomyces cerevisiae*. PLoS ONE 10(7):e0132350. https://doi.org/10.1371/journal.pone.0132350. PMID: 26147804

He M, Zhang T, Yang Y, Wang C (2021) Mechanisms of oocyte maturation and related epigenetic regulation. Front Cell Dev Biol 9:654028. https://doi.org/10.3389/fcell.2021.654028. PMID: 33842483

Horneck G, Rettberg P, Reitz G, Wehner J, Eschweiler U, Strauch K, Panitz C, Starke V, Baumstark-Khan C (2001) Protection of bacterial spores in space, a contribution to the discussion on Panspermia. Orig Life Evol Biosph 31(6):527–547. https://doi.org/10.1023/a:1012746130771. PMID: 11770260

Huang M, Hull CM (2017) Sporulation: how to survive on planet Earth (and beyond). Curr Genet 63(5):831–838. https://doi.org/10.1007/s00294-017-0694-7. PMID: 28421279

Joseph-Strauss D, Zenvirth D, Simchen G, Barkai N (2007) Spore germination in *Saccharomyces cerevisiae*: global gene expression patterns and cell cycle landmarks. Genome Biol 8(11):R241. https://doi.org/10.1186/gb-2007-8-11-r241. PMID: 17999778

Knight SJ, Goddard MR (2016) Sporulation in soil as an overwinter survival strategy in *Saccharomyces cerevisiae*. FEMS Yeast Res 16(1):fov102. https://doi.org/10.1093/femsyr/fov102. PMID: 26568201

King GA, Goodman JS, Schick JG, Chetlapalli K, Jorgens DM, McDonald KL, Ünal E (2019) Meiotic cellular rejuvenation is coupled to nuclear remodeling in budding yeast. Elife 8:e47156. https://doi.org/10.7554/eLife.47156. PMID: 31397671

King GA, Ünal E (2020) The dynamic nuclear periphery as a facilitator of gamete health and rejuvenation. Curr Genet 66(3):487–493. https://doi.org/10.1007/s00294-019-01050-1. PMID: 31915924

Koch-Bojalad BA, Carson L, Yu HG (2021) Forever young: the key to rejuvenation during gametogenesis. Curr Genet 67(2):231–235. https://doi.org/10.1007/s00294-020-01133-4. PMID: 33247310

Korza G, Setlow B, Rao L, Li Q, Setlow P (2016) Changes in bacillus spore small molecules, rRNA, germination, and outgrowth after extended sublethal exposure to various temperatures: evidence that protein synthesis is not essential for spore germination. J Bacteriol 198(24):3254–3264. https://doi.org/10.1128/JB.00583-16. PMID: 27645383

Leuschner RGK, Lillford PJ (2000) Effects of hydration on molecular mobility in phase-bright *Bacillus subtilis* spores. Microbiology (reading). 146(Pt 1):49–55. https://doi.org/10.1099/00221287-146-1-49. PMID: 10658651

Lu LY, Yu X (2015) Double-strand break repair on sex chromosomes: challenges during male meiotic prophase. Cell Cycle 14(4):516–525. https://doi.org/10.1080/15384101.2014.998070. PMID: 25565522

Maire T, Allertz T, Betjes MA, Youk H (2020) Dormancy-to-death transition in yeast spores occurs due to gradual loss of gene-expressing ability. Mol Syst Biol 16(11):e9245. https://doi.org/10.15252/msb.20199245. PMID: 33206464

Malleck T, Fekraoui F, Bornard I, Henry C, Haudebourg E, Planchon S, Broussolle V (2022) Insights into the structure and protein composition of *Moorella thermoacetica* spores formed at different temperatures. Int J Mol Sci 23(1):550. https://doi.org/10.3390/ijms23010550. PMID: 35008975

Mitton S (2022) A short history of panspermia from antiquity through the mid-1970s. Astrobiology 22(12):1379–1391. https://doi.org/10.1089/ast.2022.0032. PMID: 36475958

Mock M, Fouet A (2001) Anthrax. Annu Rev Microbiol 55:647–671. https://doi.org/10.1146/annurev.micro.55.1.647. PMID: 11544370

Moeller R, Wassmann M, Reitz G, Setlow P (2011) Effect of radioprotective agents in sporulation medium on *Bacillus subtilis* spore resistance to hydrogen peroxide, wet heat and germicidal and environmentally relevant UV radiation. J Appl Microbiol 110(6):1485–1494. https://doi.org/10.1111/j.1365-2672.2011.05004.x. PMID: 21410852

Moir A, Cooper G (2015) Spore germination. Microbiol Spectr 3(6). https://doi.org/10.1128/microbiolspec.TBS-0014-2012. PMID: 27337279

Neiman AM (2011) Sporulation in the budding yeast *Saccharomyces cerevisiae.* Genetics 189(3):737–765. https://doi.org/10.1534/genetics.111.127126. PMID: 22084423

Ohtsuka H, Imada K, Shimasaki T, Aiba H (2022) Sporulation: a response to starvation in the fission yeast *Schizosaccharomyces pombe.* Microbiologyopen 11(3):e1303. https://doi.org/10.1002/mbo3.1303. PMID: 35765188

Pechter KB, Yin L, Oda Y, Gallagher L, Yang J, Manoil C, Harwood CS (2017) Molecular basis of bacterial longevity. Mbio 8(6):e01726-e1817. https://doi.org/10.1128/mBio.01726-17. PMID: 29184015

Ray D, Ye P (2013) Characterization of the metabolic requirements in yeast meiosis. PLoS ONE 8(5):e63707. https://doi.org/10.1371/journal.pone.0063707. PMID: 23675502

Sanchez-Salas JL, Setlow B, Zhang P, Li YQ, Setlow P (2011) Maturation of released spores is necessary for acquisition of full spore heat resistance during *Bacillus subtilis* sporulation. Appl Environ Microbiol 77(19):6746–6754. https://doi.org/10.1128/AEM.05031-11. PMID: 21821751

Segev E, Smith Y, Ben-Yehuda S (2012) RNA dynamics in aging bacterial spores. Cell 148(1–2):139–149. https://doi.org/10.1016/j.cell.2011.11.059. PMID: 22209493

Setlow P (2014a) Spore resistance properties. Microbiol Spectr 2(5). https://doi.org/10.1128/microbiolspec.TBS-0003-2012. PMID: 26104355

Setlow P (2014b) Germination of spores of Bacillus species: what we know and do not know. J Bacteriol 196(7):1297–1305. https://doi.org/10.1128/JB.01455-13. PMID: 24488313

Sinclair DA, Guarente L (1997) Extrachromosomal rDNA circles–a cause of aging in yeast. Cell 91(7):1033–1042. https://doi.org/10.1016/s0092-8674(00)80493-6. PMID: 9428525

Sing TL, Brar GA, Ünal E (2022) Gametogenesis: exploring an endogenous rejuvenation program to understand cellular aging and quality control. Annu Rev Genet 56:89–112. https://doi.org/10.1146/annurev-genet-080320-025104. PMID: 35878627

Stringer JM, Winship A, Zerafa N, Wakefield M, Hutt K (2020) Oocytes can efficiently repair DNA double-strand breaks to restore genetic integrity and protect offspring health. Proc Natl Acad Sci USA 117(21):11513–11522. https://doi.org/10.1073/pnas.2001124117. PMID: 32381741

Tocheva EI, Ortega DR, Jensen GJ (2016) Sporulation, bacterial cell envelopes and the origin of life. Nat Rev Microbiol 14(8):535–542. https://doi.org/10.1038/nrmicro.2016.85. PMID: 28232669

Tu Z, Dekker HL, Roseboom W, Swarge BN, Setlow P, Brul S, Kramer G (2021) High resolution analysis of proteome dynamics during *Bacillus subtilis* sporulation. Int J Mol Sci 22(17):9345. https://doi.org/10.3390/ijms22179345. PMID: 34502250

Unal E, Kinde B, Amon A (2011) Gametogenesis eliminates age-induced cellular damage and resets life span in yeast. Science 332(6037):1554–1557. https://doi.org/10.1126/science.1204349. PMID: 21700873

Vreeland RH, Rosenzweig WD, Powers DW (2000) Isolation of a 250 million-year-old halotolerant bacterium from a primary salt crystal. Nature 407(6806):897–900. https://doi.org/10.1038/35038060. PMID: 11057666

Wang X, Montero LP (2016) Visualizing *Bacillus subtilis* during vegetative growth and spore formation. Methods Mol Biol 1431:275–287. https://doi.org/10.1007/978-1-4939-3631-1_19. PMID: 27283315

Wendland J (2020) Sporulation in *Ashbya gossypii.* J Fungi (Basel) 6(3):157. https://doi.org/10.3390/jof6030157.PMID:32872517;PMCID:PMC7558398

Wickramasinghe NC, Wickramasinghe DT, Steele EJ (2020) Cometary panspermia and origin of life? Adv Genet 106:5–20. https://doi.org/10.1016/bs.adgen.2020.03.003. PMID: 33081926

Yin L, Ma H, Nakayasu ES, Payne SH, Morris DR, Harwood CS (2019) Bacterial longevity requires protein synthesis and a stringent response. mBio 10(5):e02189-19. https://doi.org/10.1128/mBio.02189-19. PMID: 31615958

Zhang DC, Brouchkov A, Griva G, Schinner F, Margesin R (2013) Isolation and characterization of bacteria from ancient Siberian permafrost sediment. Biology (Basel) 2(1):85–106. https://doi.org/10.3390/biology2010085. PMID: 24832653

Chapter 5
Longevity Secrets of the Three Marine Beauties—Sponges, Corals, and Sea Anemones

Abstract Sponges, corals and sea anemones are ancient metazoan life forms that have much in common. They are sessile marine species of simple organization—all growing throughout their life, harboring symbiotic microorganisms, using both sexual and asexual reproduction, and lacking many regulatory and defense systems. Nevertheless, they are thriving worldwide. Sponges (*Porifera*) have only a few types of differentiated cells. Instead, they widely use microorganisms from diverse symbiotic communities. We suggest that sponges are a product of the tight symbiosis of metazoan and microbial cells, creating by essence a hybrid. In fact, they are composed partly of immortal cells of microorganisms and partly of ancient metazoan cells. Such an assembly of cells, together with an extraordinary tissue regeneration, may ensure a stunning longevity record of ca. 11,000–15,000 years! Corals and sea anemones belong to the later emerged cnidarians with a wider range of differentiated cells. Corals are able to catch some plankton and small fish, though are used to feeding via the endosymbiotic photosynthetic algae. Corals are the second in the list of the long-lived animals (over 4200 years). Sea anemones are more active and presumably thereby have only modest longevity (50–80 years).

Keywords Basal metazoans · Sponges · Corals · Sea anemones · Hydra · Dilution of waste products · Cell turnover · Tissue regeneration · Longevity · Long-lived animals

5.1 Basal Metazoans Are Flourishing Worldwide

Sponges, corals and sea anemones are marine metazoans which usually attract attention because of the contrast between the fascinating flowering and inhospitality in their environment. They occupy a phylogenetic position at the base of the metazoan tree of life, helping to understand many sophisticated evolutionary and biological relationships.

Sponges, corals and sea anemones have many common features. They are the simplest extant animals which passed successfully through evolutionary selection.

K. K. Muradian and V. E. Fraifeld, *Rejuvenation and Longevity*, Healthy Ageing and Longevity 20, https://doi.org/10.1007/978-3-031-64995-0_5

They are sessile marine species that constantly grow in size and widely use symbiosis with microorganisms. Yet, they lack many essential regulatory, repair and defense systems. Surprisingly, they are thriving worldwide—from the poles to the tropics! In the context of our story, most important is that they are record-breakers in longevity.

Biological life originated as an immortal entity. Prokaryotes and most unicellular eukaryotes are potentially immortal species. Aging emerged later as a useful adaptive process (see Chap. 2). Multicellular organisms, in contrast, are obligately mortals. When and how did this transition between immortal and mortal modes occur? The basal phylogenetic position of sponges, corals and sea anemones could shed light on these issues.

5.2 Sponges

Sponges (*Porifera*) are the ancient metazoan life forms, a phylum with over 930 species (Carrier et al. 2022; Funayama 2018). They have symmetrical and asymmetrical body shapes and grow with age, reaching over one meter in diameter. The continuous increase in body size means that they could be potential long-livers, because cellular division promotes syntheses of new macromolecules and concomitantly *dilute accumulated waste products* (see Chap. 2). Sponges are filter-feeding animals and due to the porous canal system, they pump thousands of liters of water daily, providing nutrients to the body and removing the wastes (Kiran et al. 2018; Montero-Serra et al. 2018; McGrath et al. 2018). Because of a high pumping volume, their filter cells (choanocytes) are at particular risk of injury from environmental insults. To prevent the accumulation of damage, these cells are constantly renewed at a high proliferative rate, with a duration of the cell cycle comparable to that of unicellulars (De Goeij et al. 2009). Together with cell shedding, this enables a rapid cell turnover and overall dilution of damaged cells as well as damage within the cell. Another mechanism of rapid and efficient recovery from damage in sponges is associated with their stem cell system and regenerative capacity which is the highest among metazoans (Müller 2006; Funayama 2008).

Sponges have limited types of differentiated cells (Funayama 2013). They lack the traditional animal types of cells for the guts, muscles, gills, nerves, etc. Instead, they harbor numerous microorganisms from diverse microbial communities as transient or resident endosymbionts, up to 40% of the sponge's mass, and can interact with trillions of microbes daily. Moreover, the unicellular protozoa are firmly involved in basal processes of sponges, including reproduction, nutrition, gas exchange, etc. (Hentschel et al. 2002; Sathiyanarayanan et al. 2017; Pita et al. 2016; Turon et al. 2019). This symbiosis is too complex, more resembling quasi-genetic relationships. We suggest that sponges could be a prokaryote-metazoan hybrid. They have a high proportion of totipotent stem cells, thus ensuring enhanced viability (Rando 2006; Solana 2013). Moreover, they are composed partly of the immortal symbiotic microorganisms and partly of the mortal cells that are not far differentiated yet from immortal cells. Such a mixture of 'semi-mortal' and immortal cells may lead to hardly predictable results. Then, it is not surprising that the sponges are the record-holders in longevity. The ancestors of modern sponges emerged presumably 800–900 million years ago, when Earth had extremely low oxygen content and was properly frozen (Brocks et al. 2016; Schuster et al. 2018). The existing sponges living in the deep-ocean in Antarctica and sub-Antarctic regions may practice the same attitude—the icy-cold waters with deep-sea low oxygen content and many thousand years of life span. Isotopic analysis of a deep-sea sponge *Monorhaphis chuni,* revealed a near 11,000 years of age which has been considered as a life span record for the extant animal species (Jochum et al. 2012). However, in the Guinness Book of World Records, there is a new data on Antarctic sponges presumably living for 5000–15,000 years (Abyss Ocean World 2017).

5.3 Corals

Corals are marine invertebrates of the phylum *Cnidaria.* They live worldwide in compact colonies, usually in tropical oceans at depths of 300–3000 m. The colony may create polyps of up to several meters in size. Corals are crucial for ocean biomass and biodiversity, for habituating fish and invertebrates. A coral colony consists of many genetically identical polyps. These animals have a set of tentacles surrounding the central mouth opening. Although some corals are able to catch plankton and small fish using stinging cells on their tentacles, most corals host photosynthetic endosymbiotic dinoflagellate algae (Roark et al. 2009; Bythell et al. 2018).

Coral colonies grow by asexual reproduction of polyps or breed sexually by spawning. In this case, they release gametes simultaneously overnight, usually around a full moon. Although the asexual lineage of reef corals is regarded as potentially immortal, over 80% of corals are spawners with external fertilization and larval development. While many coral species attain considerable ages (> 600 years), there are species with much shorter life span, which show a decreased reproduction, increased mortality and susceptibility to stress and diseases (Bythell et al. 2018; Ottaviani et al. 2020).

Unlike the sponges, cnidarians (corals and sea anemones included) have more types of differentiated cells and even simple diffuse networks of nerves. Their neighbors on the animal evolutionary tree, bilaterian animals, already possess a brain (Cheng 2021). Corals originated evolutionarily later and have more complex organization and daily activity than sponges. The maximum age of the two coral species in the wild, *Gerardia* sp. and *Leiopathes* sp., was estimated as 2742 years and 4265 years, respectively (Roark et al. 2009). By this data, corals are the second long-live animal species, after the sponges.

Yet, corals possess a similar to sponges regenerative capacity largely associated with participation of stem cells (Levanoni et al. 2024). Other members of the Cnidaria phylum from the genus *Hydra* are also well-known for their remarkable ability to regenerate, based mainly on a high portion of stem cells in their body. These cells are persistently and rapidly proliferating, thus enabling a high-rate cell turnover (Schaible et al. 2015; Levanoni et al. 2024). Thus, like sponges and corals, Hydras also use a mechanism of replicative dilution to minimize a damage load. Although Hydras have a relatively short life span in the wild, typically living for a few months, a several-year observation of *Hydra vulgaris* (*H. vulgaris*) in controlled laboratory conditions did not reveal any evidence for aging and senescence (Martínez 1998; Schaible et al. 2015). Moreover, it was speculated that MLS of *H. vulgaris* in favorable conditions could reach 1400 years (Jones et al. 2014). Despite such promising observations and predictions, a huge difference between the life span in the wild and in the controlled (favorable) conditions raises many still unanswered questions.

5.4 Sea Anemones

Sea anemones are also sessile marine invertebrates from the phylum *Cnidaria,* of approximately 1500 species habituating through the tropical to Sub-Antarctic and Antarctic regions (Rodríguez 2013; Fautin et al. 2013; Vassallo-Avalos et al. 2022). The anemones are experienced hunters and voracious feeders. They eat almost anything that passes by—falling mussels, shells, and small fishes or crabs, usually paralyzing the prey by the stinging cells on their tentacles, and can readily consume even polyethylene pellets (Diana et al. 2020). Sea anemones reproduce asexually (binary fission or budding) and sexually by broadcasting eggs and sperm. The largest sea anemone species in the world, *Stichodactylamertensii Brandt,* regularly exceeds one meter in oral disc diameter; it mostly reproduces via sexual reproduction and rarely reproduces asexually by budding (Bennett-Smith et al. 2022). Compared with the other basal metazoans, especially the sponges and corals, sea anemones are more active animals and have much more modest longevity records. There is some data that sea anemones can live in captivity over 50–80 years (reviewed in Petralia et al. 2014). Of note, in the last time, they are used as unique models in regenerative medicine (Amiel et al. 2021; van der Burg and Prentis 2021).

5.5 Concluding Remarks

As we already mentioned more than once, biological life originated as an immortal entity (see also Chap. 2). About three quarters of evolution occurred on the immortal prokaryotes and unicellular eukaryotes. Then, the living world became quickly occupied by mortal metazoan species. Why and when did it happen? Sponges, corals and sea anemones are at the basal phylogenetic position of metazoans and thus could shed light on this interim period between the immortal and mortal modes.

The comparison of these metazoans shows that species living in tight symbiosis with immortal microorganisms could be the record holders in longevity. They show that the transmission from the immortal unicellulars to the mortal multicellulars occurred by the principle of step by step—from the long-livers to the short-livers as their cellular organization got more complex and metabolic activity increases. Thus, the most long-lived amongst the three beauties—sponges live in tight symbiosis with immortal microorganisms, more resembling quasi-genetic relationships. Their longevity record has achieved such superiority that they seem to be living on the border between real-time mortality and fairy immortality. Corals are the second longest-live animal species (over 4200 years) after the sponges. They mostly feed by products prepared by their symbiotic partners, the photosynthetic endosymbiotic algae, though can catch some plankton and small fish. One of the practical lessons from the biology of sponges and corals is that in humans, a particular emphasis should be placed on microbiome, its interaction with the host and the ways of its correction. Another important point is their remarkable rate of cell turnover (stem cells included), which enables an efficient dilution of accumulated waste products and unprecedented tissue regeneration. Among the three beauties, the sea anemones are most active and self-sufficient in food delivery; probably because of that they have much more modest longevity records (50–80 years). That said, sea anemones have made a step from their semi-immortal 'relatives' (sponges and corals) to mortality. As noted in an elegant posit, sponges are a "connection between potentially ageless lines of single-celled organisms and the earliest and simplest multicellular animals that are equally ageless" (Petralia et al. 2014).

Ethics All the materials and data described in the book have been approved by Ethical Committee for Animal Experimentation or Helsinki Committee for Research on Humans of the organizations, where corresponding investigations have been done.

References

Abyss Ocean World (2017) Longest living sea creatures. https://abyssoceanworld.com

Amiel AR, Michel V, Carvalho JE, Shkreli M, Petit C, Röttinger E (2021) L'anémone de mer Nematostella vectensis—Un modèle émergent pour la recherche biomédicale: mécano-sensibilité, régénération et longévité [The sea anemone Nematostella vectensis, an emerging model for biomedical research: Mechano-sensitivity, extreme regeneration and longevity]. Med Sci (Paris) 37(2):167–177. French. https://doi.org/10.1051/medsci/2020282. PMID: 33591260

Bennett-Smith MF, Justo MS, Berumen ML, Peixoto R, Titus BM (2022) Novel in situ observations of asexual reproduction in the carpet sea anemone, Stichodactylamertensii (Stichodactylidae, Actiniaria). Zookeys 1103:57–63. https://doi.org/10.3897/zookeys.1103.84415. PMID: 36761782

Brocks JJ, Jarrett AJ, Sirantoine E, Kenig F, Moczydłowska M, Porter S, Hope J (2016) Early sponges and toxic protists: possible sources of cryostane, an age diagnostic biomarker antedating Sturtian Snowball Earth. Geobiology 14(2):129–149. https://doi.org/10.1111/gbi.12165. PMID: 26507690

Bythell JC, Brown BE, Kirkwood TBL (2018) Do reef corals age? Biol Rev Camb Philos Soc 93(2):1192–1202. https://doi.org/10.1111/brv.12391. PMID: 29282837

Carrier TJ, Maldonado M, Schmittmann L, Pita L, Bosch TCG, Hentschel U (2022) Symbiont transmission in marine sponges: reproduction, development, and metamorphosis. BMC Biol 20(1):100. https://doi.org/10.1186/s12915-022-01291-6. PMID: 35524305

Cheng K (2021) Learning in Cnidaria: a systematic review. Learn Behav 49(2):175–189. https://doi.org/10.3758/s13420-020-00452-3. PMID: 33439470

De Goeij JM, De Kluijver A, Van Duyl FC, Vacelet J, Wijffels RH, De Goeij AF, Cleutjens JP, Schutte B (2009) Cell kinetics of the marine sponge *Halisarca caerulea* reveal rapid cell turnover and shedding. J Exp Biol 212(Pt 23):3892–3900. https://doi.org/10.1242/jeb.034561. PMID: 19915132

Diana Z, Sawickij N, Rivera NA Jr, Hsu-Kim H, Rittschof D (2020) Plastic pellets trigger feeding responses in sea anemones. Aquat Toxicol 222:105447. https://doi.org/10.1016/j.aquatox.2020.105447. PMID: 32145485

Fautin DG, Malarky L, Soberón J (2013) Latitudinal diversity of sea anemones (Cnidaria: Actiniaria). Biol Bull 224(2):89–98. https://doi.org/10.1086/BBLv224n2p89. PMID: 23677974

Funayama N (2008) Stem cell system of sponge. In: Bosch TCG (eds) Stem cells. Springer, Dordrecht. https://doi.org/10.1007/978-1-4020-8274-0_2

Funayama N (2013) The stem cell system in demosponges: suggested involvement of two types of cells: archeocytes (active stem cells) and choanocytes (food-entrapping flagellated cells). Dev Genes Evol 223(1–2):23–38. https://doi.org/10.1007/s00427-012-0417-5. PMID: 23053625

Funayama N (2018) The cellular and molecular bases of the sponge stem cell systems underlying reproduction, homeostasis and regeneration. Int J Dev Biol 62(6–7–8):513–525. https://doi.org/10.1387/ijdb.180016nf. PMID: 29938763

Hentschel U, Hopke J, Horn M, Friedrich AB, Wagner M, Hacker J, Moore BS (2002) Molecular evidence for a uniform microbial community in sponges from different oceans. Appl Environ Microbiol 68(9):4431–4440. https://doi.org/10.1128/AEM.68.9.4431-4440.2002. PMID: 12200297

Jochum KP, Wang X, Vennemann TW, Sinha B, Müller WE (2012) Siliceous deep-sea sponge monorhaphis chuni: a potential paleoclimate archive in ancient animals. Chem Geol 300–301:143–151. https://doi.org/10.1016/j.chemgeo.2012.01.009

Jones OR, Scheuerlein A, Salguero-Gómez R, Camarda CG, Schaible R, Casper BB, Dahlgren JP, Ehrlén J, García MB, Menges ES, Quintana-Ascencio PF, Caswell H, Baudisch A, Vaupel JW (2014) Diversity of ageing across the tree of life. Nature 505(7482):169–173. https://doi.org/10.1038/nature12789. PMID: 24317695

Kiran GS, Sekar S, Ramasamy P, Thinesh T, Hassan S, Lipton AN, Ninawe AS, Selvin J (2018) Marine sponge microbial association: towards disclosing unique symbiotic interactions. Mar Environ Res 140:169–179. https://doi.org/10.1016/j.marenvres.2018.04.017. PMID: 29935729

Levanoni J, Rosner A, Lapidot Z, Paz G, Rinkevich B (2024) Coral tissue regeneration and growth is associated with the presence of stem-like cells. J Mar Sci Eng 12(2):343. https://doi.org/10.3390/jmse12020343

Martínez DE (1998) Mortality patterns suggest lack of senescence in Hydra. Experimental Gerontol 33:217–225

McGrath EC, Woods L, Jompa J, Haris A, Bell JJ (2018) Growth and longevity in giant barrel sponges: redwoods of the reef or Pines in the Indo-Pacific? Sci Rep 8(1):15317. https://doi.org/10.1038/s41598-018-33294-1.Erratum.In:SciRep.2019;9(1):18033. PMID: 30333574

Montero-Serra I, Linares C, Doak DF, Ledoux JB, Garrabou J (1873) Strong linkages between depth, longevity and demographic stability across marine sessile species. Proc Biol Sci 285(1873):20172688. https://doi.org/10.1098/rspb.2017.2688. PMID: 29491172

Müller WE (2006) The stem cell concept in sponges (Porifera): Metazoan traits. Semin Cell Dev Biol 17(4):481–491. https://doi.org/10.1016/j.semcdb.2006.05.006. PMID: 16807004

Ottaviani A, Eid R, Zoccola D, Pousse M, Dubal JM, Barajas E, Jamet K, Lebrigand K, Lapébie P, Baudoin C, Giraud-Panis MJ, Rouan A, Beauchef G, Guéré C, Vié K, Barbry P, Tambutté S, Gilson E, Allemand D (2020) Longevity strategies in response to light in the reef coral Stylophora pistillata. Sci Rep 10(1):19937. https://doi.org/10.1038/s41598-020-76925-2. PMID: 33203910

Petralia RS, Mattson MP, Yao PJ (2014) Aging and longevity in the simplest animals and the quest for immortality. Ageing Res Rev 16:66–82. https://doi.org/10.1016/j.arr.2014.05.003. PMID: 24910306

Pita L, Fraune S, Hentschel U (2016) Emerging sponge models of animal-microbe symbioses. Front Microbiol 7:2102. https://doi.org/10.3389/fmicb.2016.02102. PMID: 28066403

Rando TA (2006) Stem cells, ageing and the quest for immortality. Nature 441(7097):1080–1086. https://doi.org/10.1038/nature04958. PMID: 16810243

Roark EB, Guilderson TP, Dunbar RB, Fallon SJ, Mucciarone DA (2009) Extreme longevity in proteinaceous deep-sea corals. Proc Natl Acad Sci USA 106(13):5204–5208. https://doi.org/10.1073/pnas.0810875106. PMID: 19307564

Rodríguez E (2013) New records of Antarctic and Sub-Antarctic sea anemones (Cnidaria, Anthozoa, Actiniaria and Corallimorpharia) from the Weddell Sea, Antarctic Peninsula, and Scotia Arc. Zootaxa 3624:1–100. https://doi.org/10.11646/zootaxa.3625.1.1. PMID: 25320766

Sathiyanarayanan G, Saibaba G, Kiran GS, Yang YH, Selvin J (2017) Marine sponge-associated bacteria as a potential source for polyhydroxyalkanoates. Crit Rev Microbiol 43(3):294–312. https://doi.org/10.1080/1040841X.2016.1206060. PMID: 27824282

Schuster A, Vargas S, Knapp IS, Pomponi SA, Toonen RJ, Erpenbeck D, Wörheide G (2018) Divergence times in demosponges (Porifera): first insights from new mitogenomes and the inclusion of fossils in a birth-death clock model. BMC Evol Biol 18(1):114. https://doi.org/10.1186/s12862-018-1230-1. PMID: 30021516

Solana J (2013) Closing the circle of germline and stem cells: the primordial stem cell hypothesis. EvoDevo 4(1):2. https://doi.org/10.1186/2041-9139-4-2. PMID: 23294912

Schaible R, Scheuerlein A, Dańko MJ, Gampe J, Martínez DE, Vaupel JW (2015) Constant mortality and fertility over age in Hydra. Proc Natl Acad Sci USA 112(51):15701–15706. https://doi.org/10.1073/pnas.1521002112. PMID: 26644561

Turon M, Uriz MJ, Martin D (2019) Multipartner symbiosis across biological domains: looking at the eukaryotic associations from a microbial perspective. mSystems 4(4):e00148-19. https://doi.org/10.1128/mSystems.00148-19. PMID: 31239394

van der Burg CA, Prentis PJ (2021) The tentacular spectacular: evolution of regeneration in sea anemones. Genes (Basel) 12(7):1072. https://doi.org/10.3390/genes12071072. PMID: 34356088

Vassallo-Avalos A, Gonzlez-Muoz R, Acua FH, Cervantes-Ramrez II, Rivas G (2022) Intertidal sea anemones (Cnidaria: Actiniaria) from the west coast of the Peninsula of Baja California, Mexico. Zootaxa 5165(2):151–179. https://doi.org/10.11646/zootaxa.5165.2.1. PMID: 36095474

Chapter 6
Metazoans Who Live Longer Than Jeanne Calment

Abstract Some metazoans outlive the Methuselah's centuries while others live only few days, thereby justifying the interest in comparative longevity analysis. In usually short-livers' samplings, humans become 'comparative outliers'. To better understand the longevity-promoting factors, humans should be reconciled with species exceeding human longevity (SEHL). Among millions of metazoans, there are only 13 such species. SEHL comprise 4 fish species (Greenland shark, rockfish, sturgeon and warty oreo), 3 bivalve mollusks (ocean quahog, freshwater pearl shell and geoduck clam), 3 marine sessile species with strong symbiosis with microorganisms (sponge, coral and tubeworm), a bowhead whale, sea urchin and the only one terrestrial species giant tortoise. All, or near all, SEHL: (i) are deep- and cold-water inhabitants; (ii) they lack the mortality increase and reproduction decrease in the advanced ages; (iii) often live in hypoxic-hypercapnic environments (HHEs); (iv) preserve the life-long cell replication and body growth capacity; (v) are hypometabolic, coupled with high resistance to anabiosis. However, to become a SEHL, it is not enough to live in deep-ocean cold water. The 507 years of ocean quahogs are associated with their burrowing into the bottom stand, coupled with extreme resistance to hypoxia and anabiosis. The phenomenal longevity of Greenland sharks (392 years) may be caused by a slowing-down of deamination via reamination, mitigated by a high concentration of trimethylaminooxide. Vestimentiferan tubeworms display a low growth rate due to fairly controlled proliferation/apoptosis and kept strong symbiosis with microorganisms, eventually achieving 300 years. Bowhead whales live 211 years in HHEs and have protective adaptations against cancer and other age-related diseases. Rockfish species show amazing variability in size, habitat and longevity. Their life span increases exponentially with the ocean depth and reaches 205 years. Sea urchins exhibit negative senescence and can live 200 years without age-associated pathology. Freshwater pearl shells, habituating 190 years in the pristine Arctic rivers, remain ice-frozen in the long Polar Night winters accompanied by severe famine and anoxia. The 'longevity ikons' giant tortoises, live at least 152 years or even more. Sturgeons with MLS of 152 years belong to an ancestry dating back hundreds of million years. Geoduck clams are the largest burrowing clams in the world with MLS of at least 168 years.

K. K. Muradian and V. E. Fraifeld, *Rejuvenation and Longevity*, Healthy Ageing and Longevity 20, https://doi.org/10.1007/978-3-031-64995-0_6

Keywords Species exceeding human longevity (SEHL) · Ocean quahog · Greenland shark · Vestimentiferan tubeworms · Bowhead whales · Sea urchins · Rockfish · Freshwater pearl shells

6.1 Species Exceeding Human Longevity (SEHL)

The key objects of comparative analysis are humans who have a maximum life span (MLS) higher than the majority of animals species. Without underestimating the existing research, it should be noted that such analyses are largely focused on problems of the short-lived animals. They are more relevant for studying the factors decreasing life span, not increasing. The same logic indicates that humans should be compared with species exceeding human longevity (SEHL). Such analysis can highlight the evolutionary experience in creating species with superior anti-aging and resistance to age-related diseases (ARDs). Any knowledge on SEHL's biology could be useful for expanding the lists of experimental models and factors for health-span extension.

The search in the AnAge: The Animal Ageing and Longevity Database (Tacutu et al. 2018; https://genomics.senescence.info/species/index.html) and PubMed sources has shown that within millions of metazoan species, there are only 13 SEHL (Table 6.1). *Hydra vulgaris* (*H. vulgaris*) might be one more potential candidate to be included in the SEHL list. A several-year observation of *Hydra vulgaris (H. vulgaris)* in controlled laboratory conditions did not reveal any evidence for aging and senescence, though its predicted MLS would not exceed two and a half months (Martínez 1998; Schaible et al. 2015). According to Jones et al. (2014), an estimated (calculated) MLS of *H. vulgaris* in favorable conditions could reach 1400 years.

A quick estimate of Table 6.1 shows that all SEHL, except Giant tortoise, are aquatic animals (primary marine ones) and include 4 fish species (Greenland shark, rockfish, sturgeon and warty oreo), 3 bivalve mollusks (ocean quahog, freshwater pearl shell and geoduck clam), 3 marine sessile species with symbiosis with microorganisms (sponge, coral and vestimentiferan tubeworm) and by one species of whale and sea urchin. Although their aging mechanisms remain mostly unknown, it is clear that to live such long, SEHLs should have robust anti-aging systems, primarily against the cardiovascular, cancer and neurodegenerative pathologies.

Why among million metazoans are there only 13 SEHL? Along with specific habits, SEHL have obviously found some exceptional longevity pathways and tricks. Our task is to take lessons from these unique animals and, where possible, translate their findings into humans.

Sponges and corals, the two longest-lived species living for thousands of years, are described elsewhere in this book (see Chap. 5). We could find only very scarce data about the small fish warty oreo (*Allocyttus verrucosus*) and it was therefore excluded from this discussion. A brief description and analysis of the remaining 10 species with MLS longer than 150 years is presented in this chapter.

Table 6.1 The species exceeding human longevity (SEHL) in metazoans

MLS rating	Common name	Scientific name	Taxon	MLS, years	References
1	Sponge	*Scolymastra joubini*	Phylum Porifera	15,000	Jochum et al. (2012)
2	Corals	*Gerardia* sp. *Leiopathes* sp.	Phylum Cnidaria	2742 4265	Roark et al. (2009)
3	Ocean quahog	*Arctica islandica*	Bivalvia	507	Ungvari et al. (2013)
4	Greenland shark	*Somniosus microcephalus*	Chondrichthyes	392	Nielsen et al. (2016)
5	Vestimentiferan tubeworms	*Vestimentifera siboglinidae*	Sedentaria	300	Durkin et al. (2017)[a]
6	Bowhead whale	*Balaena mysticetus*	Mammalia	211	George et al. (1999)
7	Rockfish	*Sebastes aleutianus*	Actinopterygii	205	Munk (2001)
8	Red sea urchin	*Strongylocentrotus franciscanus*	Echinoidea	200	Ebert (2008)
9	Freshwater pearl shell	*Margaritifera margaritifera*	Bivalvia	190	Ziuganov (2004)[a]
10	Pacific geoduck	*Panopea generosa*	Bivalvia	168 179	Moss et al. (2016) Edge et al. (2021)[a]
11	Lake sturgeon	*Acipenser fulvescens*	Actinopterygii	152	Das (1994), Billard and Lecointre (2001)
12	Giant tortoise	*Geochelone gigantea*	Reptilia	152 (190)	Quesada et al. (2019)
13	Warty oreo	*Allocyttus verrucosus*	Actinopterygii	140	Cailliet et al. (2001)

[a]Not presented in AnAge

6.2 Hypoxic-Hypercapnic Environment and Anabiosis in SEHL

The metabolic rate and body temperature are the best predictors of animal longevity (Frolkis and Muradian 1991; Houtkooper et al. 2010; Lehmann et al. 2008, 2013; Protsiv et al. 2020). Oceans and particularly their polar regions are cozy homes for the hypometabolic and low-body-temperature species because the O_2 content and environmental temperature decrease as the marine depth increases. Moreover, calorie restriction during the long polar nights in Antarctic and Arctic regions could be an additional longevity-promoting factor.

Though there are few direct experimental data, it is obvious that at least one third of SEHL (4 of 13 species) live not only in hypoxic but also in hypercapnic environment (HHE). This is the case in the three species of mollusks presented in Table 6.1 (Ocean quahog, Pacific geoduck, Freshwater pearl shell), which most time are burrowed in the ground's sands or muds. They close the channels of gaseous exchange and this results in an increase in Pco_2 and decrease in Po_2 up to the state close to anabiosis. HHE is also inevitable in deep-diving whales, who can remain under the water for about one hour. The above species clearly demonstrate that low body Po_2 and high Pco_2 are not obligatory associated with pathological conditions but can be achieved via numerous and well-regulated physiological systems, e.g., used by the species with negligible senescence (Tolstun et al. 2020; Fraifeld and Muradian 2021; Muradian and Fraifeld 2021).

The concise analysis revealed that all, or nearly all, SEHL (i) are deep- and cold-water inhabitants, (ii) often living in hypoxic-hypercapnic environment (HHE), (iii) are hypometabolic and highly resistant to anabiosis; (iv) they lack the age-dependent increase in mortality and decrease in reproduction; (v) preserve life-long cell replication and body-grow capacity. Of note, most of these features are typical for negligible and negative senescence as well.

6.3 Ocean Quahog

Ocean quahogs (*Arctica islandica*) with MLS of 507 years (Ungvari et al. 2013) are in the third place in the list of the longest-lived metazoans (Table 6.1). They are the leader of the triple group of the longest-lived bivalve mollusks in the SEHL list, followed by freshwater pearl shells and Pacific geoducks. Ocean quahogs demonstrate exclusively low metabolic rate and high resistance to anoxia. In free laboratory conditions, they can remain burrowed up to 6 days at the partial pressure of O_2 (Po_2) close to zero (Strahl et al. 2011). In forced conditions of anoxia, they show just stunning results: on the 55th day of the exposure to anoxia, 50% of animals were still alive (Philipp et al. 2012).

Although eukaryote tree has a variety of mitochondrial inheritance pathways, metazoans adhere mostly the pattern of the maternal lineage inheritance. The latter reduces the risk of mitonuclear genomic incompatibility, decreases aging rate and associated pathology (Latorre-Pellicer et al. 2016). Ocean quahogs often use the doubly uniparental inheritance (DUI)—the natural occurrence of maternal and parental types of mitochondrial genomes. Such a phenomenon, when both types of mitochondrial genes are transcribed in somatic cells, provokes substantial drop of the respiratory activity (Dégletagne et al. 2021), which could be an additional factor to decrease the metabolic rate. Study of aerobic capacity in the mitochondria of ocean quahogs in short-term hypoxia (24 h, $< 0.01\%$ O_2) and subsequent reoxygenation (1.5 h, 21% O_2) further revealed a unique resistance of these animals to both O_2 deficiency and O_2 toxic effects. It was found that the substrate oxidation and proton leak subsystems in ocean quahog mitochondria were not affected by hypoxia and reoxygenation. Indeed, the ocean quahog mitochondria show a robust response to hypoxia by maintaining stable substrate oxidation and upregulating phosphorylation capacity (Steffen et al. 2021).

Currently, fully sequenced mtDNA and data on its base composition are available for 7 SEHL, ocean quahogs included (Table 6.2). Remarkably, these SEHL use another than mammals mtDNA longevity strategy (Fig. 6.1 and Table 6.3). In opposite to mammalian species which longevity is to a great extent determined by GC content and associated thermodynamic stability of mtDNA (Lehmann et al. 2008, 2013; Toren et al. 2016), the long-lived Bivalvia and SEHL in general exhibit a high mtDNA AT/GC ratio. It means that in SEHL mtDMA prevail double hydrogen bonds, formed by AT, over the triple hydrogen bonds in GC pairs. This allows for quick unfolding of double stranded mtDNA during its replication, transcription and repair. Of note, in contrast to GC, AT pairs are less vulnerable for spontaneous deamination because A and T do not have amino groups in positions which are sensitive to mutations.

Cardiac aging is crucial for most animal species. In ocean quahogs, no significant age-related changes in the caspase-like and trypsin-like proteasome activity, oxidative protein damage, protein ubiquitination, and expression of heat shock proteins, inflammatory factors, and mitochondrial electron transport chain complexes in the heart were observed (Sosnowska et al. 2014).

Table 6.2 Base contents of mtDNA heavy strand in SEHL

Name	Species	GC%	AT%	AT/GC	G%	C%	A%	T%
Ocean quahog	*Artica islandica*	30.7	69.3	2.25	10.8	19.9	40.3	29.0
Pacific geoduck	*Panopea generosa*	36.2	63.7	1.76	25.0	11.2	25.0	38.7
Freshwater pearl shell	*Margaritifera margaritifera*	39.3	60.7	1.54	25.8	13.6	23.4	37.3
Greenland shark	*Somniosus microcephalus*	39.3	60.7	1.54	14.4	24.84	30.8	29.91
Rockfish	*Sebastes aleutianus*	45.6	54.4	1.19	17.4	28.15	27.5	26.94
Sturgeon	*Acipenser fulvescens*	45.8	54.2	1.18	16.5	29.32	30.0	24.14
Bowhead whale	*Balaena mysticetus*	41.3	58.7	1.42	28.1	13.2	26.0	32.7
Humans	*Homo sapience*	44.4	55.6	1.25	31.3	13.1	24.7	30.9

Data on human mtDNA is also included

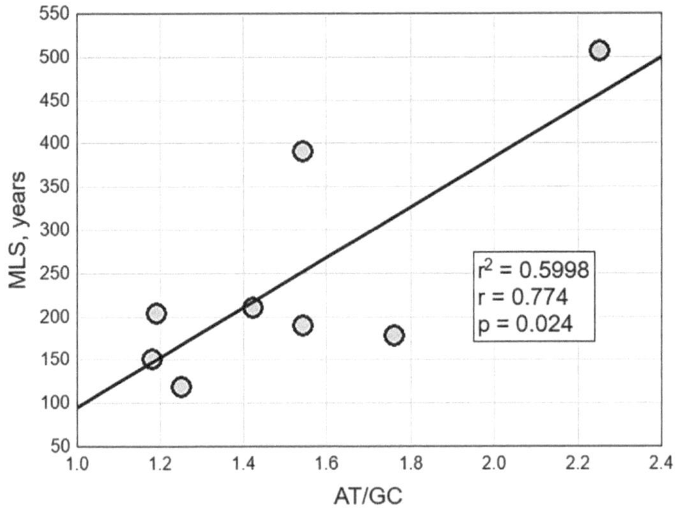

Fig. 6.1 Longevity (MLS) correlates with mtDNA AT/GC ratio in SEHL (n = 8)

Table 6.3 Correlations of mtDNA base composition (heavy strand) with longevity in SEHL

Name	r^2	r	*p*-value
AT%	0.562	0.75	0.032
GC%	0.559	0.74	0.032
AT/GC	0.851	0.77	0.024
A%	0.751	0.86	0.008
G%	0.56	−0.75	0.044
T%	0.029	−0.17	> 0.7
C%	0.047	−0.21	> 0.6

Data on human mtDNA is also included

6.4 Greenland Shark

The Greenland sharks (*Somniosus microcephalus,* GSs) are iconic predators living in the cold deep waters of the North Atlantic. The largest recorded specimen weights over 1000 kg and has length around 6.5 m (Nielsen et al. 2020). In natural ecological conditions, the Greenland sharks display a very low metabolism which corresponds to only 60–190 g of fish daily (Ste-Marie et al. 2022). Its low metabolic rate is supported by the low aortic blood pressure and slow-speed swimming (Shadwick et al. 2018). In contrast to the vast majority of fish species, the Greenland sharks are viviparous. They give birth to about 10 embryos with length of 40 cm (Carter and Soma 2020; Nielsen et al. 2020). Radiocarbon dating of eye lens nuclei showed that the oldest Greenland shark could be the longest-lived vertebrate (392 ± 120 years) (Nielsen et al. 2016).

The flesh of Greenland sharks is toxic to most animals and humans. Analysis of the toxic fractions indicates the presence of large amounts of trimethylamine oxide (TMAO) but no other substances could be directly responsible for the poisoning. The symptoms appear to be due to an acute trimethylamine (TMA) poisoning arising from intestinal reduction of TMAO to TMA (Anthoni et al. 1991). Damage caused by spontaneous deamination of nucleic acids and proteins is one of the oldest hypotheses of aging (Gromenko et al. 2009). We assume that in Greenland shark, deamination could be counterbalanced by *spontaneous reamination* mitigated by elevated amino-containing substances like TMAO. Consequently, it is reasonable to suggest that the extraordinary longevity of Greenland sharks is largely attributed to a high level of spontaneous reamination.

6.5 Vestimentiferan Tubeworms

Vestimentiferan tubeworms are one of the longest-lived marine invertebrates. These worm-like sessile invertebrates have neither a mouth nor a stomach and do not eat. Instead, billions of symbiotic bacteria living inside the tubeworms produce some sugars from H_2S and O_2 spewed from the benthic oceanic vents. In this regard, the vestimentiferan tubeworms resemble, to a great extent, the longest-lived animals—sponges and corals who live in tight symbiosis with immortal microorganisms (see Chap. 5). Tubeworms can stabilize sediments and generate oases for benthic communities. They form symbiosis with sulfur-oxidizing bacteria which are mostly localized in the cells of a highly vascularized tissue of the worm. The high molecular hemoglobin of the worms is used as a transporter for vents' H_2S and O_2. Symbiotic bacteria in the tubeworm use the vents' H_2S as an energy source for themselves and for the worms (Minic and Hervé 2004; Callaway et al. 2010; Thornburg et al. 2010).

Direct sperm transfer from males to females has been documented in some tubeworms species while other species discharge primary oocytes. Since brooding of tubeworm embryos has never been observed, it suggests that fertilization might be external. Yet, in five species of tubeworms, sperm was storage at the posterior end of the oviduct, where eggs are inseminated internally, fertilized and meiosis completed after eggs were released from the female (Hilário et al. 2005). Surprisingly, vestimentiferan tubeworms living around short-term hydrothermal vents grow faster and live shorter, while the worms habituating at the long-persistent cold vents can live hundreds of years. Although the basic mechanisms of such adaptation remain unknown, it should be noted that apoptosis is down-regulated in vestimentiferan tubeworms with quick proliferation, in contrast to the balanced proliferation/apoptosis in slowly aging tubeworms (Pflugfelder et al. 2009; Durkin et al. 2017).

6.6 Bowhead Whales

Due to the extraordinary land-to-sea transition, whales became fully aquatic and acquired unique subcutaneous adipose layers, the blubber. It can reach up to 50 cm in depth and compose around 40–50% of the 90,000 kg total body mass in mature adults. It is multifunctional and serves as metabolic reservoir for lipid deposition and utilization, for synthesis of physiological regulators such as hormones and mediators, provides buoyancy and serves as an insulator for thermoregulation, etc. These giants feed mostly marine invertebrates, krill and small fish (Ball et al. 2015, 2017). Bowhead whales confer resistance to lipid oxidation, presumably due to a high sphingolipid content (Borchman et al. 2017). Bowhead whales apparently possess protective mechanisms against age-related diseases, cancer in particular (Prokopov 2007). The analysis of its transcriptomics revealed longevity adaptations associated with DNA maintenance and repair, cell growth (cell death and cell-cycle regulation), ubiquitination, immune responses and dietary adaptations (Seim et al. 2014; Keane et al. 2015; Toren et al. 2020).

As we mentioned above, the metabolic rate and body temperature are the best predictors of animal longevity. For obvious reasons, it is hardly possible to evaluate the metabolic rate in such a big animal as a bowhead whale. Yet, its body temperature (33.6 °C) is one of the lowest across the mammals (Prokopov 2007). The mitochondrial DNA (mtDNA) GC content is another major predictor of mammalian longevity (Lehmann et al. 2008, 2013). The higher mtDNA GC content, the higher thermodynamic stability of mitochondrial genome. This could be particularly important for endotherms. Combination of low body temperature with a relatively high mtDNA GC content (see Table 6.2) could contribute to extreme longevity of the bowhead whales among warm-blooded animals.

Aspartic acid L/D racemization in the eye lens nucleus showed that MLS of bowhead whales could be at least 211 years, making them the longest-lived mammalian species (George et al. 1999). Their MLS is almost twice longer than that of humans and places the bowhead whales in the sixth place among the longest-lived metazoans.

6.7 Sea Urchins

Around a thousand species represent sea urchins as marine invertebrates in the class *Echinoidea*. Due to the sharp teeth, they eat anything that floats by and due to the delicate flesh they are favorite preys for many predators and humans. Sea urchins show indeterminate growth and lack of elevated mortality with age. They can maintain regenerative abilities throughout life. Both the short-lived and long-lived species demonstrate negative senescence manifesting in an increase in reproduction capacity, and a decrease in mortality rate with advanced age (Ebert 2019; Amir et al. 2020). Sea urchins do not show age-related telomere shortening and a decreased telomerase activity in somatic tissues (Kogan et al. 2015). They generate millions of eggs and sperm, alas, getting back only a handful to grow into adults. The red sea urchin, *Mesocentrotus franciscanus*, is in the eighth position among SEHL, with MLS of 200 years (Francis et al. 2006; Ebert 2008, 2019; Amir et al. 2020).

RNA-Seq analysis of transcriptional profiles in the muscle and esophagus tissues of sea urchins revealed only a few age-related changes in genes expression (Polinski et al. 2020). Some age-related changes were found in expression of protein-coding genes involved in the ubiquitin–proteasome system, translational regulation, and electron transport chain.

6.8 Rockfish

Rockfish encompasses over a hundred of species of the genus *Sebastes*, with an amazing variability in the size, habitat, longevity, etc. They inhabit the northeast Pacific Ocean and elsewhere and their MLS ranges within 10–205 years (Munk 2001; Love et al. 2002; Mangel et al. 2007). There are several rockfish species with MLS over 100 years, which successfully resist deleterious effects of age—epigenetic drift, telomere shortening and genome instability, oxidative damage and proteostasis collapse (Heras and Aguilar 2019; Treaster et al. 2023). Age validation of rockfish species showed that their longevity increases exponentially with depth (Cailliet et al. 2001). Rockfishes are a good model for analysis of aging and longevity because they have stable rates of growth and fecundity, negligible senescence and a long life span. The evolution of rockfish couples the genetic diversity with reshaping of mutational spectrum in segregating CpG → TpG variants in nuclear genome of long-lived species (Kolora et al. 2021; Treaster et al. 2023). Comparative analysis showed that rockfishes life span correlated negatively with the rate of mtDNA mutations, but not with the rate of nuclear DNA mutations (Hua et al. 2015). The nuclear mutations are primarily linked to defects in the DNA replication in the germline cells. Then, the longevity of rockfish species could be related to mitochondria rather than nuclear genome.

6.9 Freshwater Pearl Shell

Freshwater pearl shells (*Margaritifera margaritifera*) are the ninth longest-lived metazoans. Like other bivalve mollusks, the age rings laid down every year in the shell are reliable measures of their longevity. Comparison of six populations of freshwater pearl shells from Spain and three Arctic populations showed that the southern species have MLS within 28–40 years, while in Arctic rivers MLS is much higher, ranging from 114 to 190 years. Northern populations of freshwater pearl shells mostly live in small rivers that can freeze in the long winter months. The Polar Night winters are times of low metabolic rates and severe calorie restriction, up to famine and anabiosis.

Freshwater pearl shells often burrow in the sediment and close tightly their valves. This results in a dramatic drop in tissue Po_2, up to anoxia. Tissue repair and defense mechanisms against diseases (cancer, infections, etc.) seem to be extraordinarily high in Freshwater pearl shells, especially in the cold climates, thus contributing to their longevity (Philipp and Abele 2010). Indeed, around 40% of mussels lived more than 100 years (Ziuganov et al. 2000, Ziuganov 2004).

6.10 Geoduck Clams

The Pacific geoduck clams (*Panopea generosa*) are one of the largest burrowing clams in the world. The oldest geoducks can reach 168 (Moss et al. 2016) or even 179 years (Edge et al. 2021). Pre-exposure of geoducks to moderate environmental challenges enhances their ability to cope with subsequent severe chronic or repeated stress insults (Gurr et al. 2021). The experiments on *Panopea generosa* clams that were exposed to high levels of pCO_2 in seawater, showed that pre-acclimated animals activated genes involved in homeostatic processes, protein degradation, stress, apoptosis, innate immune responses, and mitochondrial recycling (Gurr et al. 2022). Interestingly, the Pacific geoduck mitochondrial genome has a high AT content (Bisbal-Pardo et al. 2016), which is a second after *Arctica islandica* (Table 6.2; see also section "Ocean quahog").

6.11 Giant Tortoise

Tortoises are long-livers and endangered reptiles of the family *Testudinidae*. Their 152 years of MLS only slightly exceed the SEHL's lower boarder (122.5 years). However, the Guinness World Records book identified a new 190-year-old tortoise as the oldest one, thus significantly increasing its prestige as a long-lived species. A study of over 50 turtles and tortoises living in captivity (zoos and aquariums) revealed that approximately 75% of species showed slow or negligible senescence (da Silva et al. 2022). Global comparative analysis of the genomes of giant tortoise and related species revealed lineage-specific variants of genes involved in DNA repair, inflammatory responses and cancer development, thus indicating specific evolutionary strategies for life span extension (Krivoruchko and Storey 2010; Quesada et al. 2019). Of note, the same biological categories were found as longevity adaptations in long-lived mammalian species including the species with negligible senescence (Seim et al. 2014; Keane et al. 2015; Toren et al. 2020).

6.12 Sturgeon

Sturgeons are a very old fish lineages with an ancestry dating back hundreds million years. They show indeterminate growth and very slow or negligible senescence (Patnaik et al. 1994). Sturgeons and *Acipenser fulvescens* in particular exhibit late sexual maturity (15–17 years in males and 20–24 years in females) and long period between two spawning (four years on average) (Billard and Lecointre 2001). Most sturgeon species are vulnerable to overfishing and contamination, and thus are endangered (Hung 2017; Degani et al. 2022). Lower environmental temperature and scarcity of food significantly retard senescence and promote their longevity. Since sturgeons grow through their entire life, the largest fish usually means the oldest one. MLS of sturgeons reaches 152 years. Of note, like the sturgeons, Australian lungfishes, *Neoceratodus forsteri*, with MLS close to 100 years, also remain relatively unchanged during their hundreds million-year evolutionary history (Das 1994; Patnaik et al. 1994).

6.13 Concluding Remarks

Metazoans' life span varies in a wide range. This provides the opportunity for comparative analysis of longevity and extreme longevity in particular. The key object of such kind of research are humans who have a relatively long life span and often become outliers when compared with the shorter-lived animals. However, these studies are more relevant for searching the factors that decrease longevity rather than those that promote longevity. Otherwise, humans should be analyzed in 'company' with SEHL in an attempt to clarify what makes a potentially long-lived animal an extraordinarily long-livers. We have found only 13 such species which exceed human longevity. Unfortunately, our knowledge of the biology of these exotic animals is far from being sufficiently full. Nevertheless, the analysis of SEHL revealed a number of conditions and factors that could contribute to their extreme life span. First, all SEHL live in an intermittent (bowhead whales) or permanent (most other SEHL) hypoxic environment often combined with hypercapnia. Second, the vast majority of SEHL are cold-blooded animals living in cold water, thereby having a low body temperature and hypometabolism. The only one warm-blooded SEHL, the bowhead whale, exhibits one of the lowest body temperatures among the mammals. Third, many SEHL experience long-term periods of starvation. Fourth, apart from environmental conditions, SEHL possess some specific mechanisms and features that could promote longevity, among them, spontaneous reamination of nucleic acids and proteins (Greenland sharks), prevalence of double hydrogen bonds in mtDNA (bivalves, deep-water fishes, giant tortoises), the aforementioned low body temperature and hypometabolism. Besides, SEHL preserve the growth capacity throughout the entire life span. Some SEHL (sponges, corals, tubeworms) live in tight symbiosis with immortal microorganisms which take on the main metabolic burden.

Study in depth of the SEHL biology could bring about exciting discoveries for our understanding of basic mechanisms of longevity and rejuvenation.

Ethics All the materials and data described in the book have been approved by Ethical Committee for Animal Experimentation or Helsinki Committee for Research on Humans of the organizations, where corresponding investigations have been done.

References

Amir Y, Insler M, Giller A, Gutman D, Atzmon G (2020) Senescence and longevity of sea urchins. Genes (Basel) 11(5):573. https://doi.org/10.3390/genes11050573. PMID: 32443861

Anthoni U, Christophersen C, Gram L, Nielsen NH, Nielsen P (1991) Poisonings from flesh of the Greenland shark *Somniosus microcephalus* may be due to trimethylamine. Toxicon 29(10):1205–1212. https://doi.org/10.1016/0041-0101(91)90193-u. PMID: 1801314

Ball HC, Stavarz M, Oldaker J, Usip S, Londraville RL, George JC, Thewissen JG, Duff RJ (2015) Seasonal and ontogenetic variation in subcutaneous adipose of the bowhead whale (*Balaena mysticetus*). Anat Rec (Hoboken) 298(8):1416–1423. https://doi.org/10.1002/ar.23125. PMID: 25711480

Ball HC, Londraville RL, Prokop JW, George JC, Suydam RS, Vinyard C, Thewissen JG, Duff RJ (2017) Beyond thermoregulation: metabolic function of cetacean blubber in migrating bowhead and beluga whales. J Comp Physiol B 187(1):235–252. https://doi.org/10.1007/s00360-016-1029-6. PMID: 27573204

Billard R, Lecointre G (2001) Biology and conservation of sturgeon and paddlefish. Rev Fish Biol Fish 10:355e92

Bisbal-Pardo CI, Del Río-Portilla MA, Rocha-Olivares A (2016) The complete mitochondrial DNA of the Pacific Geoduck clam (*Panopea generosa*). Mitochondrial DNA A DNA Mapp Seq Anal 27(3):1955–1956. https://doi.org/10.3109/19401736.2014.971304. PMID: 25329256

Borchman D, Stimmelmayr R, George JC (2017) Whales, lifespan, phospholipids, and cataracts. J Lipid Res 58(12):2289–2298. https://doi.org/10.1194/jlr.M079368. PMID: 29038122

Callaway R, Desroy N, Dubois SF, Fournier J, Frost M, Godet L, Hendrick VJ, Rabaut M (2010) Ephemeral bio-engineers or reef-building polychaetes: how stable are aggregations of the tube worm *Lanice conchilega* (Pallas, 1766)? Integr Comp Biol 50(2):237–250. https://doi.org/10.1093/icb/icq060. PMID: 21558202

Cailliet G, Andrews A, Burton E et al (2001) Age determination and validation studies of marine fishes: do deep-dwellers live longer? Exp Gerontol 36:739–764. https://doi.org/10.1016/S0531-5565(00)00239-4. PMID: 11295512

Carter AM, Soma H (2020) Viviparity in the longest-living vertebrate, the Greenland shark (*Somniosus microcephalus*). Placenta 97:26–28. https://doi.org/10.1016/j.placenta.2020.05.014. PMID: 32792058

Das M (1994) Age determination and longevity in fishes. Gerontology 40(2–4):70–96. https://doi.org/10.1159/000213580. PMID: 7926859

da Silva R, Conde DA, Baudisch A, Colchero F (2022) Slow and negligible senescence among testudines challenges evolutionary theories of senescence. Science 376(6600):1466–1470. https://doi.org/10.1126/science.abl7811. PMID: 35737795

Degani G, Nevo Sarel M, Hajouj A, Hurvitz A, Veksler-Lublinsky I, Meerson A (2022) Whole-genome inter-sex variation in Russian sturgeon (*Acipenser gueldenstaedtii*). Int J Mol Sci 23(16):9469. https://doi.org/10.3390/ijms23169469. PMID: 36012734

Dégletagne C, Abele D, Glöckner G, Alric B, Gruber H, Held C (2021) Presence of male mitochondria in somatic tissues and their functional importance at the whole animal level in the marine bivalve *Arctica islandica*. Commun Biol 4(1):1104. https://doi.org/10.1038/s42003-021-02593-1. PMID: 34545198

Durkin A, Fisher CR, Cordes EE (2017) Extreme longevity in a deep-sea vestimentiferan tubeworm and its implications for the evolution of life history strategies. Naturwissenschaften 104(7–8):63. https://doi.org/10.1007/s00114-017-1479-z. PMID: 28689349

Ebert TA (2008) Longevity and lack of senescence in the red sea urchin *Strongylocentrotus franciscanus*. Exp Gerontol 43(8):734–738. https://doi.org/10.1016/j.exger.2008.04.015. PMID: 18550313

Ebert TA (2019) Negative senescence in sea urchins. Exp Gerontol 122:92–98. https://doi.org/10.1016/j.exger.2019.04.018. PMID: 31063808

Edge D, Reynolds D, Wanamaker A, Griffin D, Bureau D, Outridge C, Wang R, Stevick B, Black B (2021) Multicentennial proxy record of northeast pacific sea surface temperatures from the annual growth increments of *Panopea generosa*. Paleoceanogr Paleoclimatol 36(9). Bibcode:2021PaPa...36.4291E. https://doi.org/10.1029/2021PA004291. S2CID 239151578

Fraifeld VE, Muradian KK (2021) Hypoxic-hypercapnic environment as a model of hypometabolism, calorie restriction, hypoglycemia and non-medicamentous treatment of related pathology. Diabetes Res Metab 2(1):110

Francis N, Gregg T, Owen R, Ebert T, Bodnar A (2006) Lack of age-associated telomere shortening in long- and short-lived species of sea urchins. FEBS Lett 580(19):4713–4717. https://doi.org/10.1016/j.febslet.2006.07.049. PMID: 16876792

Frolkis VV, Muradian KhK (1991) Life span prolongation. CRC Press, Boca Raton

George JC, Bada J, Zeh J, Scott L, Brown SE, O'Hara T, Suydam R (1999) Age and growth estimates of bowhead whales (*Balaena mysticetus*) via aspartic acid racemization. Can J Zool 77:571–580

Gromenko EV, Spirin PV, Kubareva EA, Romanova EA, Prassolov VS, Shpanchenko OV, Dontsova OA (2009) Deamination of 5-methylcytosine residues in Mammalian cells. Acta Naturae. 1(3):121–124 PMID: 22649624

Gurr SJ, Wanamaker SA, Vadopalas B, Roberts SB, Putnam HM (2021) Repeat exposure to hypercapnic seawater modifies growth and oxidative status in a tolerant burrowing clam. J Exp Biol 224(13):jeb233932. https://doi.org/10.1242/jeb.233932. PMID: 34027545

Gurr SJ, Trigg SA, Vadopalas B, Roberts SB, Putnam HM (2022) Acclimatory gene expression of primed clams enhances robustness to elevated pCO_2. Mol Ecol 31(19):5005–5023. https://doi.org/10.1111/mec.16644. PMID: 35947503

Heras J, Aguilar A (2019) Comparative transcriptomics reveals patterns of adaptive evolution associated with depth and age within marine rockfishes (*Sebastes*). J Hered 110(3):340–350. https://doi.org/10.1093/jhered/esy070. PMID: 30602025

Hilário A, Young CM, Tyler PA (2005) Sperm storage, internal fertilization, and embryonic dispersal in vent and seep tubeworms (Polychaeta: Siboglinidae: Vestimentifera). Biol Bull 208(1):20–28. https://doi.org/10.2307/3593097. PMID: 15713809

Houtkooper RH, Williams RW, Auwerx J (2010) Metabolic networks of longevity. Cell 142(1):9–14. https://doi.org/10.1016/j.cell.2010.06.029. PMID: 20603007

Hua X, Cowman P, Warren D, Bromham L (2015) Longevity is linked to mitochondrial mutation rates in rockfish: a test using Poisson regression. Mol Biol Evol 32(10):2633–2645. https://doi.org/10.1093/molbev/msv137. PMID: 26048547

Hung SSO (2017) Recent advances in sturgeon nutrition. Anim Nutr 3(3):191–204. https://doi.org/10.1016/j.aninu.2017.05.005. PMID: 29767119

Jochum KP, Wang X, Vennemann TW, Sinha B, Müller WE (2012) Siliceous deep-sea sponge *Monorhaphis chuni*: a potential paleoclimate archive in ancient animals. Chem Geol 300:143–151

Jones OR, Scheuerlein A, Salguero-Gómez R, Camarda CG, Schaible R, Casper BB, Dahlgren JP, Ehrlén J, García MB, Menges ES, Quintana-Ascencio PF, Caswell H, Baudisch A, Vaupel JW (2014) Diversity of ageing across the tree of life. Nature 505(7482):169–173. https://doi.org/10.1038/nature12789. PMID: 24317695

Keane M, Semeiks J, Webb AE, Li YI, Quesada V, Craig T, Madsen LB, van Dam S, Brawand D, Marques PI, Michalak P, Kang L, Bhak J, Yim HS, Grishin NV, Nielsen NH, Heide-Jørgensen MP, Oziolor EM, Matson CW, Church GM, Stuart GW, Patton JC, George JC, Suydam R, Larsen K, López-Otín C, O'Connell MJ, Bickham JW, Thomsen B, de Magalhães JP (2015) Insights into the evolution of longevity from the bowhead whale genome. Cell Rep 10(1):112–122. https://doi.org/10.1016/j.celrep.2014.12.008. PMID: 25565328

Kogan V, Molodtsov I, Menshikov LI, Shmookler Reis RJ, Fedichev P (2015) Stability analysis of a model gene network links aging, stress resistance, and negligible senescence. Sci Rep 5:13589. https://doi.org/10.1038/srep13589. PMID: 26316217

Kolora SRR, Owens GL, Vazquez JM, Stubbs A, Chatla K, Jainese C, Seeto K, McCrea M, Sandel MW, Vianna JA, Maslenikov K, Bachtrog D, Orr JW, Love M, Sudmant PH (2021) Origins and evolution of extreme life span in Pacific Ocean rockfishes. Science 374(6569):842–847. https://doi.org/10.1126/science.abg5332. PMID: 34762458

Krivoruchko A, Storey KB (2010) Forever young: mechanisms of natural anoxia tolerance and potential links to longevity. Oxid Med Cell Longev 3(3):186–198. https://doi.org/10.4161/oxim.3.3.12356. PMID: 20716943

Latorre-Pellicer A, Moreno-Loshuertos R, Lechuga-Vieco AV, Sánchez-Cabo F, Torroja C, Acín-Pérez R, Calvo E, Aix E, González-Guerra A, Logan A, Bernad-Miana ML, Romanos E, Cruz R, Cogliati S, Sobrino B, Carracedo Á, Pérez-Martos A, Fernández-Silva P, Ruíz-Cabello J, Murphy MP, Flores I, Vázquez J, Enríquez JA (2016) Mitochondrial and nuclear DNA matching shapes metabolism and healthy ageing. Nature 535(7613):561–565. https://doi.org/10.1038/nature18618. PMID: 27383793

Lehmann G, Segal E, Muradian KK, Fraifeld VE (2008) Do mitochondrial DNA and metabolic rate complement each other in determination of the mammalian maximum longevity? Rejuvenation Res 11(2):409–417. https://doi.org/10.1089/rej.2008.0676. PMID: 18442324

Lehmann G, Muradian KK, Fraifeld VE (2013) Telomere length and body temperature-independent determinants of mammalian longevity? Front Genet 4:111. https://doi.org/10.3389/fgene.2013.00111. PMID: 23781235

Love MS, Yoklavich M, Thorsteinson L (2002) The rockfishes of the northeast Pacific. University of California Press, Berkeley, CA

Mangel M, Kindsvater HK, Bonsall MB (2007) Evolutionary analysis of life span, competition, and adaptive radiation, motivated by the Pacific rockfishes (*Sebastes*). Evolution 61(5):1208–1224. https://doi.org/10.1111/j.1558-5646.2007.00094.x. PMID: 17492972

Martínez DE (1998) Mortality patterns suggest lack of senescence in Hydra. Experimental Gerontol 33:217–225. PMID: 9615920

Minic Z, Hervé G (2004) Biochemical and enzymological aspects of the symbiosis between the deep-sea tubeworm *Riftia pachyptila* and its bacterial endosymbiont. Eur J Biochem 271(15):3093–3102. https://doi.org/10.1111/j.1432-1033.2004.04248.x. PMID: 15265029

Moss DK, Ivany LC, Judd EJ, Cummings PW, Bearden CE, Kim WJ, Artruc EG, Driscoll JR (1836) Lifespan, growth rate, and body size across latitude in marine Bivalvia, with implications for Phanerozoic evolution. Proc Biol Sci 283(1836):20161364. https://doi.org/10.1098/rspb.2016.1364. PMID: 27488653

Munk KM (2001) Maximum ages of groundfishes in waters of Alaska and British Columbia and considerations of age determination. Alsk Fish Res Bull 8:12–21

Muradian KK, Fraifeld VE (2021) Hypercapnia-inducible factor: a hypothesis. Aging & Longevity (a&l). 2(3):27–31

Nielsen J, Hedeholm RB, Heinemeier J, Bushnell PG, Christiansen JS, Olsen J, Ramsey CB, Brill RW, Simon M, Steffensen KF, Steffensen JF (2016) Eye lens radiocarbon reveals centuries of longevity in the Greenland shark (*Somniosus microcephalus*). Science 353(6300):702–704. https://doi.org/10.1126/science.aaf1703. PMID: 27516602

Nielsen J, Hedeholm RB, Lynghammar A, McClusky LM, Berland B, Steffensen JF, Christiansen JS (2020) Assessing the reproductive biology of the Greenland shark (*Somniosus microcephalus*). PLoS ONE 15(10):e0238986. https://doi.org/10.1371/journal.pone.0238986. PMID: 33027263

Patnaik BK, Mahapatro N, Jena BS (1994) Ageing in fishes. Gerontology 40(2–4):113–132. https://doi.org/10.1159/000213582. PMID: 7926851

Pflugfelder B, Cary SC, Bright M (2009) Dynamics of cell proliferation and apoptosis reflect different life strategies in hydrothermal vent and cold seep vestimentiferan tubeworms. Cell Tissue Res 337(1):149–165. https://doi.org/10.1007/s00441-009-0811-0. PMID: 19444472

Philipp EE, Abele D (2010) Masters of longevity: lessons from long-lived bivalves–a mini-review. Gerontology 56(1):55–65. https://doi.org/10.1159/000221004. PMID: 19468199

Philipp EE, Wessels W, Gruber H, Strahl J, Wagner AE, Ernst IM, Rimbach G, Kraemer L, Schreiber S, Abele D, Rosenstiel P (2012) Gene expression and physiological changes of different populations of the long-lived bivalve *Arctica islandica* under low oxygen conditions. PLoS ONE 7(9):e44621. https://doi.org/10.1371/journal.pone.0044621. PMID: 23028566

Polinski JM, Kron N, Smith DR, Bodnar AG (2020) Unique age-related transcriptional signature in the nervous system of the long-lived red sea urchin *Mesocentrotus franciscanus*. Sci Rep 10(1):9182. https://doi.org/10.1038/s41598-020-66052-3. PMID: 32514014

Prokopov AF (2007) Theoretical paper: exploring overlooked natural mitochondria-rejuvenative intervention: the puzzle of bowhead whales and naked mole rats. Rejuvenation Res 10(4):543–560. https://doi.org/10.1089/rej.2007.0546. PMID: 18072884

Protsiv M, Ley C, Lankester J, Hastie T, Parsonnet J (2020) Decreasing human body temperature in the United States since the industrial revolution. Elife 9:e49555. https://doi.org/10.7554/eLife.49555. PMID: 31908267

Quesada V, Freitas-Rodríguez S, Miller J, Pérez-Silva JG, Jiang ZF, Tapia W, Santiago-Fernández O, Campos-Iglesias D, Kuderna LFK, Quinzin M, Álvarez MG, Carrero D, Beheregaray LB, Gibbs JP, Chiari Y, Glaberman S, Ciofi C, Araujo-Voces M, Mayoral P, Arango JR, Tamargo-Gómez I, Roiz-Valle D, Pascual-Torner M, Evans BR, Edwards DL, Garrick RC, Russello MA, Poulakakis N, Gaughran SJ, Rueda DO, Bretones G, Marquès-Bonet T, White KP, Caccone A, López-Otín C (2019) Giant tortoise genomes provide insights into longevity and age-related disease. Nat Ecol Evol 3(1):87–95. https://doi.org/10.1038/s41559-018-0733-x. PMID: 30510174

Roark EB, Guilderson TP, Dunbar RB, Fallon SJ, Mucciarone DA (2009) Extreme longevity in proteinaceous deep-sea corals. Proc Natl Acad Sci USA 106(13):5204–5208. https://doi.org/10.1073/pnas.0810875106. PMID: 19307564

Schaible R, Scheuerlein A, Dańko MJ, Gampe J, Martínez DE, Vaupel JW (2015) Constant mortality and fertility over age in Hydra. Proc Natl Acad Sci USA 112(51):15701–15706. https://doi.org/10.1073/pnas.1521002112. PMID: 26644561

Seim I, Ma S, Zhou X, Gerashchenko MV, Lee SG, Suydam R, George JC, Bickham JW, Gladyshev VN (2014) The transcriptome of the bowhead whale *Balaena mysticetus* reveals adaptations of the longest-lived mammal. Aging (Albany NY) 6(10):879–899. https://doi.org/10.18632/aging.100699. PMID: 25411232

Shadwick RE, Bernal D, Bushnell PG, Steffensen JF (2018) Blood pressure in the Greenland shark as estimated from ventral aortic elasticity. J Exp Biol 221(Pt 19):jeb186957. https://doi.org/10.1242/jeb.186957. PMID: 30104302

Sosnowska D, Richardson C, Sonntag WE, Csiszar A, Ungvari Z, Ridgway I (2014) A heart that beats for 500 years: age-related changes in cardiac proteasome activity, oxidative protein damage and expression of heat shock proteins, inflammatory factors, and mitochondrial complexes in *Arctica islandica*, the longest-living noncolonial animal. J Gerontol A Biol Sci Med Sci 69(12):1448–1461. https://doi.org/10.1093/gerona/glt201. PMID: 24347613

Steffen JBM, Haider F, Sokolov EP, Bock C, Sokolova IM. Mitochondrial capacity and reactive oxygen species production during hypoxia and reoxygenation in the ocean quahog, *Arctica islandica*. J Exp Biol 224(21):jeb243082. https://doi.org/10.1242/jeb.243082. PMID: 34697625

Ste-Marie E, Watanabe YY, Semmens JM, Marcoux M, Hussey NE. Life in the slow lane: field metabolic rate and prey consumption rate of the Greenland shark (*Somniosus microcephalus*) modelled using archival biologgers. J Exp Biol 225(7):jeb242994. https://doi.org/10.1242/jeb.242994. PMID: 35258589

Strahl J, Brey T, Philipp EE, Thorarinsdóttir G, Fischer N, Wessels W, Abele D (2011) Physiological responses to self-induced burrowing and metabolic rate depression in the ocean quahog *Arctica islandica*. J Exp Biol 214(Pt 24):4223–4233. https://doi.org/10.1242/jeb.055178. PMID: 22116766

Tacutu R, Thornton D, Johnson E, Budovsky A, Barardo D, Craig T, Diana E, Lehmann G, Toren D, Wang J, Fraifeld VE, de Magalhães JP (2018) Human ageing genomic resources: new and updated databases. Nucleic Acids Res 46(D1):D1083–D1090. https://doi.org/10.1093/nar/gkx1042. PMID: 29121237

Thornburg CC, Zabriskie TM, McPhail KL (2010) Deep-sea hydrothermal vents: potential hot spots for natural products discovery? J Nat Prod 73(3):489–499. https://doi.org/10.1021/np900662k. PMID: 20099811

Tolstun DA, Knyazer A, Tushynska TV, Dubiley TA, Bezrukov VV, Fraifeld VE, Muradian KK (2020) Metabolic remodelling of mice by hypoxic-hypercapnic environment: imitating the naked mole-rat. Biogerontology 21(2):143–153. https://doi.org/10.1007/s10522-019-09848-9. PMID: 31667660

Toren D, Barzilay T, Tacutu R, Lehmann G, Muradian KK, Fraifeld VE (2016) MitoAge: a database for comparative analysis of mitochondrial DNA, with a special focus on animal longevity. Nucleic Acids Res 44(D1):D1262–D1265. https://doi.org/10.1093/nar/gkv1187. PMID: 26590258

Toren D, Kulaga A, Jethva M, Rubin E, Snezhkina AV, Kudryavtseva AV, Nowicki D, Tacutu R, Moskalev AA, Fraifeld VE (2020) Gray whale transcriptome reveals longevity adaptations associated with DNA repair and ubiquitination. Aging Cell 19(7):e13158. https://doi.org/10.1111/acel.13158. PMID: 32515539

Treaster S, Deelen J, Daane JM, Murabito J, Karasik D, Harris MP (2023) Convergent genomics of longevity in rockfishes highlights the genetics of human life span variation. Sci Adv 9(2):eadd2743. https://doi.org/10.1126/sciadv.add2743. PMID: 36630509

Ungvari Z, Sosnowska D, Mason JB, Gruber H, Lee SW, Schwartz TS, Brown MK, Storm NJ, Fortney K, Sowa J, Byrne AB, Kurz T, Levy E, Sonntag WE, Austad SN, Csiszar A, Ridgway I (2013) Resistance to genotoxic stresses in *Arctica islandica*, the longest living noncolonial animal: is extreme longevity associated with a multistress resistance phenotype? J Gerontol A Biol Sci Med Sci 68(5):521–529. https://doi.org/10.1093/gerona/gls193. PMID: 23051979

Ziuganov VV (2004) Arctic and southern freshwater pearl mussel *Margaritifera margaritifera* with long and short life span as a model system for testing longevity mechanisms. Adv Gerontol 14:21–30 PMID: 15559495

Ziuganov V, Miguel ES, Neves RJ, Longa A, Fernandez C, Amaro R, Beletsky V, Popkovitch E, Kaliuzhin S, Johnson T (2000) Life span variation of the freshwater pearl shell: a model species for testing longevity mechanisms in animals. Ambio 29:102–105

Chapter 7
Transposable Elements and Rejuvenation of Genome

Abstract Transposable elements (TEs) are ubiquitous relicts of ancient viral invasions. Due to the mobility and virus-adopted viability, retroviruses became among the most abundant coding sequences in nature and the prominent force of evolution. TEs can generate new genes or change the expression of neighboring genes. They play a role in rejuvenative totipotency in early embryogenesis and pluripotency of inner cell mass (ICM) in pre-implanted blastocyst. 'Domesticated' retrotransposons can be co-opted in conflicts against invasive agents, including alien TEs. Retroviruses have a long history, apparently with periods of activation and recession. Comparative analysis of mammalian species shows that the rate of endogenous retroviruses (ERVs) integration in the human genome and the genome of other hominoids experienced a several fold decline during the last 10 million years. Using this ratio of new/old integrations, we observed a significant correlation with species maximum life span. The negative relation ($r = -0.663$; $p < 10^{-5}$) indicates that long-lived mammals and especially humans exhibit elevated protection against new events of viral integration. Retrovirus can bring new information and co-opt other TEs for utilization of new functionality. Consequently, nearly 70% of the human genome is derived from retroviruses. It provokes a ridiculous question: are humans the mammals with the majority of genes of viral origin, or are they successors of viruses with the minority of mammalian genes? TEs have a controversial nature. They could be usefully symbiotic in embryogenesis, stringently silent in mid-adulthood and loosely controlled and destructive in aging. It is believed that cumulative effects of even moderate errors in cellular structure and function can cause genomic instability and less reliable TEs silencing. Is there anything forcing TEs to act destructively in aging, or is it the virus-inherited 'instincts?' After all, death of a host is counter-productive for TEs too. Without deciphering the role of TEs in embryogenesis and aging, discussion of human rejuvenation seems illusive.

Keywords Dormant transposable elements · Active transposable elements · LINE1 · Early embryonic development · Zygotic genome activation · Endogenous retroviruses · Mammalian longevity

K. K. Muradian and V. E. Fraifeld, *Rejuvenation and Longevity*, Healthy Ageing and Longevity 20, https://doi.org/10.1007/978-3-031-64995-0_7

7.1 In 'Stormy Sea' of Transposable Elements

Transposable elements (TEs) are ubiquitous mobile genetic sequences with controversial effects. With few exceptions, they have been found in the genomes of all tissues and species of plant and metazoan origin (Huang et al. 2012; Larouche et al. 2020). Retroviruses have repeatedly invaded and successfully proliferated in the mammalian genome during the last 150 million years, including invasions in primate species during the last 80 million years (Cordaux and Batzer 2009; Richardson et al. 2015). The vast majority of TEs are dormant and lack regulatory or executive mechanisms for genome expansion or horizontal transfer (Chuong et al. 2017; Platt et al. 2018). Nevertheless, TEs engage host genomes in long 'arms races', thus activating defense systems (Nishihara 2020). Host cells employ multiple and stage-specific mechanisms to silence TEs. Epigenetic defense, histone modifications and DNA methylation, play a key role in TEs silencing in both somatic and germ cells, whereas base pairing of endogenous retroviruses (ERVs) with small non-coding PIWI-interacting RNA (piRNAs) can suppress mammalian germline TEs (Yang and Wang 2016). In *Hydra*, the PIWI-piRNA pathway suppresses TEs in somatic stem cells as well, which may contribute to lack of senescence in this organism (Teefy et al. 2020).

These inherently mutagenic viral elements can change their location in the genome, strongly prohibited for host sequences. Due to the mobility and virus-adopted viability, retroviruses are ubiquitous. They are among the most abundant coding sequences in nature and are considered as the prominent forces of evolution (Hughes and Coffin 2001; Hayward and Gilbert 2022). Often named by negative terms 'parasitic DNA', 'selfish genetic elements' or 'jumping genes', TEs made a real 'jumping ahead' to restore their reputation.

TEs can generate new genes or change the expression of neighboring genes up to reprogramming human networks. The 'domesticated' retrotransposons can be co-opted for conflicts against invasive agents, including viruses and alien TEs (Jangam et al. 2017; Cosby et al. 2019). Although many details remain unknown, it has become increasingly clear that *TEs are involved in rejuvenation events of early development—in establishing the totipotency in early embryos, its transition to pluripotency and 'last minute' correction of ICM cells in pre-implanted blastocyst* (Göke et al. 2015; Durruthy-Durruthy et al. 2016; Percharde et al. 2018; Muradian and Fraifeld 2019; Low et al. 2021). At the same time, TEs are considered to be among vicious factors in cancer, ARDs and aging (Hancks and Kazazian 2016; Gorbunova et al. 2021; Brochard et al. 2023; Liu et al. 2023). TEs and ERVs in particular, being mostly dormant in postnatal development and young ages, become awakened again after mid-adulthood, thus contributing to the senescence of somatic cells and tissue aging. Their activation was observed in organs of aged primates and mice as well as in tissues and serum from the elderly humans (Liu et al. 2023). Altogether, this demonstrates the *antagonistic pleiotropic nature of TEs* that being beneficial and indispensable in early development, become harmful later in life (Frolkis and Muradian 1991).

The evolutionary success of TEs can hardly be explained by the notorious retrovirus aggression. Rather, TEs have applied a symbiosis-like strategy, mitigating the cost of their propagation. Can TEs be involved in cellular computation and decision-making (see Chap. 3)? Will millions of years of 'close cooperation' with the human genome ensure their symbiotic behavior? Anyway, the formerly dominating genomic research on RNA- or protein-coding sequences which make only 1.5% of the DNA looks like a small island in 'stormy sea' of TEs.

7.2 We Have So Much in Common …

Retrovirus invaders could present an important source of novelty and co-opted other TEs for performance of new functions (Almeida et al. 2022). Consequently, nearly half of the mammalian genome is believed to be derived from retroviruses (Mager and Stoye 2015). Because ancient genetic insertions usually become unidentifiable by the homology-based methods, the more accurate estimates show that TEs content may reach even more, making up stunning 70% of the human genome (de Koning et al. 2011). This encourages to take a closer look at ourselves. Speaking in legal terms, are humans the mammals with majority of viral genes or successors of viruses with minority of mammalian genes?! ☺

7.3 Active and Dormant Transposable Elements

There are only a few documented horizontal transfers in mammalian history, so that TEs spread mostly by vertical inheritance (Peccoud et al. 2017). Depending on the mode of the replication, TEs are divided into two classes: RNA-based retroviruses (RVs) that use the 'copy and paste' mechanism, and DNA transposons applying the 'cut and paste' machinery. Typical mammalian genome has nearly 3% of DNA transposons which have lost their mobility (except in bats) (Campos-Sánchez et al. 2014). Retrotransposons of long interspersed elements (LINEs) and short interspersed elements (SINEs) are the mammalian main TEs encompassing, correspondingly, 17% and 11% of the human genome. The mouse genome also contains active long terminal repeat (LTR)-containing retrotransposons that are inactive in humans (Hancks and Kazazian 2016).

7.4 Transposable Elements in the First Meeting of Rejuvenation and Aging

The maternal-to-zygotic transition, also known as zygotic genome activation (ZGA), is the essence of *embryogenic rejuvenation* (see also Chap. 11). The largest and the smallest cells in mammalian species, the ovum and the spermatozoid, get united in a fertilized egg warranted to create a new organism. In terms of mortality and longevity, a spermatozoid and ovum are ordinary cells with a restricted lifespan. However, these 'mortal' cells are mysteriously transformed into an immortal totipotent zygote, an event which occurs with direct participation of TEs! As such, it could be the dramatic though not fully conceivable story of the short meeting and departure of rejuvenation and aging in mammals. After the first specific divisions of zygote, totipotency is replaced by pluripotency. Again, like during ZGA, this is accompanied by elevation in transcriptional activity of specific ERVs in the ICM cells of pre-implanted blastocyst (Low et al. 2021).

7.5 LINE1 and Other Retrotransposons in Early Embryonic Development

The members of LINE1 family are the largest and the only autonomous retrotransposons. It constitutes 17% of the human genome and 19% of the mouse genome. However, in mice, among 900,000 LINE1 sequences approximately 3000 are still active (Hancks and Kazazian 2016; Garcia-Perez et al. 2016), while in human genome, only a small subset of 80–100 sequences of total 500,000 LINE1 could be active (Brouha et al. 2003).

Inhibition of reverse transcriptase by specific drugs or microinjection of specific antibodies caused early embryonic developmental arrest. Microinjection of antisense targeting LINE1 retrotransposons in murine zygotes irreversibly arrested embryo development at the two- and four-cell stages, supporting the assumption that expression of LINE1 is inevitably required for early embryonic development (Beraldi et al. 2006; Percharde et al. 2018). Overall, elevation in transcriptional activity of specific retrotransposons/ERVs is evident as early as in ZGA and is maintained up to the stage of pre-implanted blastocyst (Low et al. 2021). Retrotransposons/ERVs were shown to be important in preservation of pluripotency of ESCs (Robbez-Masson and Rowe 2015).

7.6 Zygotic Genome Activation

Zygotic genome activation proceeds through two main phases—minor ZGA and major ZGA (Abe et al. 2018). In mice, the minor ZGA occurs between S phase of the one-cell embryo and G1 of the two-cell embryo, while the major ZGA occurs between mid-two-cell stage and late-two-cell stage. During minor ZGA, starts the heterochromatin unwinding and initiation of the first transcription reactions followed by the burst of transcriptional activity during the major ZGA. Taking into account that the sites of heterochromatic relaxation for initiation of transcription could be localized in various parts of the genome, mobility of ERVs could give them advantages to fulfill the task. We assume that *ERVs could be a candidate in creating mobile vehicles for opening of the transcription cites in ZGA*. Epigenetic resetting of the parental genomes initiates selective activation of zygotic genes, including a great number of ERVs and retrotransposons (Macfarlan et al. 2012; Göke et al. 2015; Low et al. 2021).

7.7 The New/Old Integrations of Endogenous Retroviruses and Mammalian Longevity

Endogenous retroviruses (ERVs) are obvious remnants of ancient retroviral infections which found 'cozy' habitat and became an incorporated part of the host genome (Blikstad et al. 2008). They have a long history, with apparent periods of activation and recession. Despite significant species-specific differences, past and recent evolution of ERVs remain poorly understood. Comparative analysis of 40 mammalian species showed that ERVs integration in humans and other hominoids experienced a several fold decline during the last 10 million years (Magiorkinis et al. 2015). Using these data, we found a highly significant negative correlation ($r = -0.663$; $p < 10^{-5}$) between the ratio of new/old ERVs integrations and mammalian longevity (Fig. 7.1).

This observation indicates that the long-lived mammals are more sustainable to incorporation of new ERVs than the short-lived ones. It could be said that the long-lived mammalian species have actually established the host-ERVs symbiotic relationships, while short-lived species have not yet reached this state and continue to

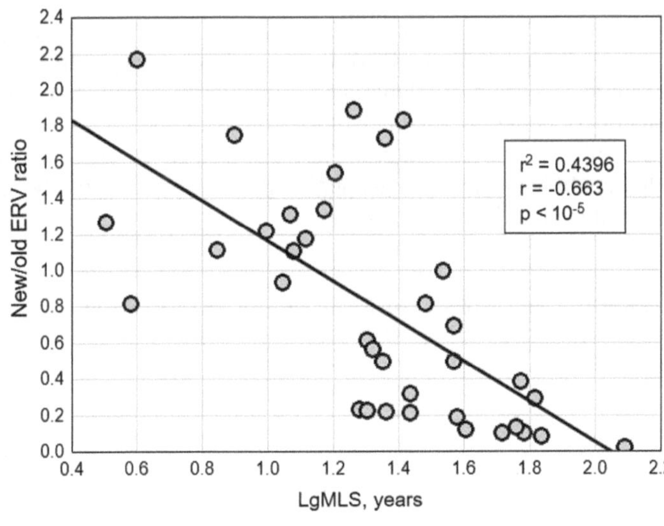

Fig. 7.1 Correlation of the ratio of new/old ERVs with MLS in mammals. MLS records were taken from AnAge database: https://genomics.senescence.info/species/index.html; Tacutu et al. (2018)

elevate the ERVs number (Fig. 7.1). In light of the above, it does not seem paradoxical another finding showing that the TE abundancy displays a significant positive correlation with mammalian longevity (paper in preparation).

The human genome appeared especially stable to integration of new ERVs. There have been only a few invasions during the last 10 million years of evolution. The same trend—a dramatic decrease in new ERVs insertions—was observed in long-lived vs short-lived species of bats, as well as in the longest-lived rodent naked mole-rat vs short-lived rodent species (Ricci et al. 2023).

7.8 Concluding Remarks

Transposable elements represent a prominent example of *antagonistic pleiotropy*: they are indispensable in early embryonic development, being involved in establishing the zygote totipotency and pluripotency of ESCs but become harmful later in life. Transcriptional activity of specific TEs/ERVs coincides well with the critical time-points of embryonic rejuvenation: first, during ZGA and establishing the totipotency, and the second, during 'last minute' correction of pluripotent ICM cells in preimplantation blastocyst (see also Chap. 11).

The evolutionary success of TEs can hardly be explained by the notorious retroviral aggression. Rather, they have applied a symbiotic strategy. It appears that the most ancient TEs lost their 'aggressive' features along with the negative effects in advanced ages. Only 'young', relatively newly acquired TEs can be destructive in aging. Consequently, we observed a dramatic decrease in new ERVs insertions in the genome of long-lived mammals.

The most prominent evolutionary achievement of TEs is their occupation of around 70% of the human genome. In view of a widely recognized ability of DNA to function as a huge system of interconnected nanoprocessors (see Chap. 3), TEs offer the potential of being involved in cellular computation.

TEs have another enigmatic feature. They can find an envelope and leave the genome. What could happen with rejuvenation if this process would be stimulated by environmental factors or genome editing?

Ethics All the materials and data described in the book have been approved by Ethical Committee for Animal Experimentation or Helsinki Committee for Research on Humans of the organizations, where corresponding investigations have been done.

References

Abe KI, Funaya S, Tsukioka D, Kawamura M, Suzuki Y, Suzuki MG, Schultz RM, Aoki F (2018) Minor zygotic gene activation is essential for mouse preimplantation development. Proc

Natl Acad Sci USA 115(29):E6780–E6788. https://doi.org/10.1073/pnas.1804309115. PMID: 29967139

Almeida MV, Vernaz G, Putman ALK, Miska EA (2022) Taming transposable elements in vertebrates: from epigenetic silencing to domestication. Trends Genet 38(6):529–553. https://doi.org/10.1016/j.tig.2022.02.009. PMID: 35307201

Beraldi R, Pittoggi C, Sciamanna I, Mattei E, Spadafora C (2006) Expression of LINE-1 retroposons is essential for murine preimplantation development. Mol Reprod Dev 73(3):279–287. https://doi.org/10.1002/mrd.20423. PMID: 16365895

Blikstad V, Benachenhou F, Sperber GO, Blomberg J (2008) Evolution of human endogenous retroviral sequences: a conceptual account. Cell Mol Life Sci 65(21):3348–3365. https://doi.org/10.1007/s00018-008-8495-2. PMID: 18818874

Brochard T, McIntyre RL, Houtkooper RH, Seluanov A, Gorbunova V, Janssens GE (2023) Repurposing nucleoside reverse transcriptase inhibitors (NRTIs) to slow aging. Ageing Res Rev 92:102132. https://doi.org/10.1016/j.arr.2023.102132. PMID: 37984625

Brouha B, Schustak J, Badge RM, Lutz-Prigge S, Farley AH, Moran JV, Kazazian HH Jr (2003) Hot L1s account for the bulk of retrotransposition in the human population. Proc Natl Acad Sci USA 100(9):5280–5285. https://doi.org/10.1073/pnas.0831042100. PMID: 12682288

Campos-Sánchez R, Kapusta A, Feschotte C, Chiaromonte F, Makova KD (2014) Genomic landscape of human, bat, and ex vivo DNA transposon integrations. Mol Biol Evol 31(7):1816–1832. https://doi.org/10.1093/molbev/msu138. PMID: 24809961

Chuong EB, Elde NC, Feschotte C (2017) Regulatory activities of transposable elements: from conflicts to benefits. Nat Rev Genet 18(2):71–86. https://doi.org/10.1038/nrg.2016.139. PMID: 27867194

Cosby RL, Chang NC, Feschotte C (2019) Host-transposon interactions: conflict, cooperation, and cooption. Genes Dev 33(17–18):1098–1116. https://doi.org/10.1101/gad.327312.119. PMID: 31481535

Cordaux R, Batzer MA (2009) The impact of retrotransposons on human genome evolution. Nat Rev Genet 10(10):691–703. https://doi.org/10.1038/nrg2640. PMID: 19763152

de Koning AP, Gu W, Castoe TA, Batzer MA, Pollock DD (2011) Repetitive elements may comprise over two-thirds of the human genome. PLoS Genet 7(12):e1002384. https://doi.org/10.1371/journal.pgen.1002384. PMID: 22144907

Durruthy-Durruthy J, Sebastiano V, Wossidlo M, Cepeda D, Cui J, Grow EJ, Davila J, Mall M, Wong WH, Wysocka J, Au KF, Reijo Pera RA (2016) The primate-specific noncoding RNA HPAT5 regulates pluripotency during human preimplantation development and nuclear reprogramming. Nat Genet 48(1):44–52. https://doi.org/10.1038/ng.3449. PMID: 26595768

Frolkis VV, Muradian KK (1991) Life span prolongation. CRC Press, Boca Raton, p 425

Garcia-Perez JL, Widmann TJ, Adams IR (2016) The impact of transposable elements on mammalian development. Development 143(22):4101–4114. https://doi.org/10.1242/dev.132639. PMID: 27875251

Göke J, Lu X, Chan YS, Ng HH, Ly LH, Sachs F, Szczerbinska I (2015) Dynamic transcription of distinct classes of endogenous retroviral elements marks specific populations of early human embryonic cells. Cell Stem Cell 16(2):135–141. https://doi.org/10.1016/j.stem.2015.01.005. PMID: 25658370

Gorbunova V, Seluanov A, Mita P, McKerrow W, Fenyö D, Boeke JD, Linker SB, Gage FH, Kreiling JA, Petrashen AP, Woodham TA, Taylor JR, Helfand SL, Sedivy JM (2021) The role of retrotransposable elements in ageing and age-associated diseases. Nature 596(7870):43–53. https://doi.org/10.1038/s41586-021-03542-y. PMID: 34349292

Hancks DC, Kazazian HH Jr (2016) Roles for retrotransposon insertions in human disease. Mob DNA 7:9. https://doi.org/10.1186/s13100-016-0065-9. PMID: 27158268

Hayward A, Gilbert C (2022) Transposable elements. Curr Biol 32(17):R904–R909. https://doi.org/10.1016/j.cub.2022.07.044. PMID: 36099891

Huang CR, Burns KH, Boeke JD (2012) Active transposition in genomes. Annu Rev Genet 46:651–675. https://doi.org/10.1146/annurev-genet-110711-155616. PMID: 23145912

Hughes JF, Coffin JM (2001) Evidence for genomic rearrangements mediated by human endogenous retroviruses during primate evolution. Nat Genet 29(4):487–489. https://doi.org/10.1038/ng775. PMID: 11704760

Jangam D, Feschotte C, Betrán E (2017) Transposable element domestication as an adaptation to evolutionary conflicts. Trends Genet 33(11):817–831. https://doi.org/10.1016/j.tig.2017.07.011. PMID: 28844698

Larouche JD, Trofimov A, Hesnard L, Ehx G, Zhao Q, Vincent K, Durette C, Gendron P, Laverdure JP, Bonneil É, Côté C, Lemieux S, Thibault P, Perreault C (2020) Widespread and tissue-specific expression of endogenous retroelements in human somatic tissues. Genome Med 12(1):40. https://doi.org/10.1186/s13073-020-00740-7. PMID: 32345368

Liu X, Liu Z, Wu Z, Ren J, Fan Y, Sun L, Cao G, Niu Y, Zhang B, Ji Q, Jiang X, Wang C, Wang Q, Ji Z, Li L, Esteban CR, Yan K, Li W, Cai Y, Wang S, Zheng A, Zhang YE, Tan S, Cai Y, Song M, Lu F, Tang F, Ji W, Zhou Q, Belmonte JCI, Zhang W, Qu J, Liu GH (2023) Resurrection of endogenous retroviruses during aging reinforces senescence. Cell 186(2):287-304.e26. https://doi.org/10.1016/j.cell.2022.12.017. PMID: 36610399

Low Y, Tan DEK, Hu Z, Tan SYX, Tee WW (2021) Transposable element dynamics and regulation during zygotic genome activation in mammalian embryos and embryonic stem cell model systems. Stem Cells Int 2021:1624669. https://doi.org/10.1155/2021/1624669. PMID: 34691189

Macfarlan TS, Gifford WD, Driscoll S, Lettieri K, Rowe HM, Bonanomi D, Firth A, Singer O, Trono D, Pfaff SL (2012) Embryonic stem cell potency fluctuates with endogenous retrovirus activity. Nature 487(7405):57–63. https://doi.org/10.1038/nature11244.PMID:22722858;PMCID:PMC3395470

Mager DL, Stoye JP (2015) Mammalian endogenous retroviruses. Microbiol Spectr 3(1):MDNA3-0009-2014. https://doi.org/10.1128/microbiolspec.MDNA3-0009-2014. PMID: 26104559

Magiorkinis G, Blanco-Melo D, Belshaw R (2015) The decline of human endogenous retroviruses: extinction and survival. Retrovirology 12:8. https://doi.org/10.1186/s12977-015-0136-x. PMID: 25640971

Muradian KK, Fraifeld VE (2019) iPSCs—induced cellular reprogramming. In: Rattan S (ed) Encyclopedia of biomedical gerontology, , vol 2. Elsevier, pp 291–298

Nishihara H (2020) Transposable elements as genetic accelerators of evolution: contribution to genome size, gene regulatory network rewiring and morphological innovation. Genes Genet Syst 94(6):269–281. https://doi.org/10.1266/ggs.19-00029. PMID: 31932541

Peccoud J, Loiseau V, Cordaux R, Gilbert C (2017) Massive horizontal transfer of transposable elements in insects. Proc Natl Acad Sci USA 114(18):4721–4726. https://doi.org/10.1073/pnas.1621178114. PMID: 28416702

Percharde M, Lin CJ, Yin Y, Guan J, Peixoto GA, Bulut-Karslioglu A, Biechele S, Huang B, Shen X, Ramalho-Santos M (2018) A LINE1-nucleolin partnership regulates early development and ESC identity. Cell 174(2):391-405.e19. https://doi.org/10.1016/j.cell.2018.05.043. PMID: 29937225

Platt RN 2nd, Vandewege MW, Ray DA (2018) Mammalian transposable elements and their impacts on genome evolution. Chromosome Res 26(1–2):25–43. https://doi.org/10.1007/s10577-017-9570-z. PMID: 29392473

Ricci M, Peona V, Boattini A, Taccioli C (2023) Comparative analysis of bats and rodents' genomes suggests a relation between non-LTR retrotransposons, cancer incidence, and ageing. Sci Rep 13(1):9039. https://doi.org/10.1038/s41598-023-36006-6. PMID: 37270634

Richardson SR, Doucet AJ, Kopera HC, Moldovan JB, Garcia-Perez JL, Moran JV (2015) The influence of LINE-1 and SINE retrotransposons on mammalian genomes. Microbiol Spectr 3(2):MDNA3-0061-2014. https://doi.org/10.1128/microbiolspec.MDNA3-0061-2014. PMID: 26104698

Robbez-Masson L, Rowe HM (2015) Retrotransposons shape species-specific embryonic stem cell gene expression. Retrovirology 12:45. https://doi.org/10.1186/s12977-015-0173-5. PMID: 26021318

Tacutu R, Thornton D, Johnson E, Budovsky A, Barardo D, Craig T, Diana E, Lehmann G, Toren D, Wang J, Fraifeld VE, de Magalhães JP (2018) Human ageing genomic resources: new and updated databases. Nucleic Acids Res 46(D1):D1083–D1090. https://doi.org/10.1093/nar/gkx1042. PMID: 29121237

Teefy BB, Siebert S, Cazet JF, Lin H, Juliano CE (2020) PIWI-piRNA pathway-mediated transposable element repression in Hydra somatic stem cells. RNA 26(5):550–563. https://doi.org/10.1261/rna.072835.119. PMID: 32075940

Yang F, Wang PJ (2016) Multiple LINEs of retrotransposon silencing mechanisms in the mammalian germline. Semin Cell Dev Biol 59:118–125. https://doi.org/10.1016/j.semcdb.2016.03.001. PMID: 26957474

Chapter 8
Embryonic Stem Cells and the 'Last-Minute' Correction

Abstract Embryonic stem cells (ESCs) are the totipotent/pluripotent stem cells which initiate formation of a mammalian body through over 220 cells of inner cell mass (ICM) of the preimplanted blastocyst in the oviduct. We assume that the number of ICM cells defines the variety of cell types in a multicellular organism by the principle 'one ICM cell—one type of a differentiated cell'. The preliminarily estimates show that the development and multiplication of these cells in humans and long-lived mammalian species occur in severe hypoxic-hypercapnic environment (HHE). HHE may determine the rate of metabolic processes and proliferation of ESCs. The severity of Po_2 in oviduct strongly correlates with mammalian longevity. Also, the duration of ESC exposure to HHE (i.e., a period from zygote to ICM of the blastocyst prior to its implantation) correlates with species-specific MLS.

Keywords Embryonic stem cells (ESCs) · Cellular pluripotency · Hypoxic-hypercapnic environment · Mammalian embryogenesis · ESCs growth rate · Mammalian longevity · ICM cells

8.1 Embryonic Stem Cells as a Gold Standard of Cellular Pluripotency and Immortality

Over 40 years ago, embryonic stem cells (ESCs) have been discovered in mice (mESCs) (Evans and Kaufman 1981; Martin 1981) and 17 years later they have been found in humans (hESCs) (Thomson et al. 1998). These natural stem cells have many opportunities for cellular therapy. Yet, their wide application has been limited because of the biomedical (tumor or immune complications) and ethical issues as well as technical restraints (Yamanaka 2020; Pendse et al. 2022). There are several ways that to some extent may by-pass the unwanted effects. For example, ESC-derived extracellular vesicles can ensure partial rejuvenation of senescent cells and antagonize aging in mice in vivo and in vitro (Yu et al. 2023; Rudnitsky et al. 2024). Anyway, ESCs hold a great promise in rejuvenation and anti-aging, primarily,

K. K. Muradian and V. E. Fraifeld, *Rejuvenation and Longevity*, Healthy Ageing and Longevity 20, https://doi.org/10.1007/978-3-031-64995-0_8

because they are pluripotent stem cells with unlimited self-renewal and malleable genetic editing (Hanna et al. 2010; Horii and Hatada 2015; Zhang et al. 2023).

8.2 Hypoxic-Hypercapnic Environment (HHE) in Mammalian Embryogenesis

Every new human life begins in HHE. The partial pressure of O_2 (Po_2) is significantly lower and the partial pressure of CO_2 (Pco_2) is higher in the embryo and fetus compared with neonatal and adult organisms. Moreover, the longer-lived mammalian species start their life in more severe HHE than the short-lived ones (Muradian and Fraifeld 2023). For example, in humans and rhesus monkeys, Po_2 in the oviduct varied around 1% and Pco_2 reached 12–19% (Maas et al. 1977; Fischer and Bavister 1993; Ng et al. 2018). Our analysis of currently available data on mammalian species with reported values of Po_2 in the oviduct demonstrated that the development of ESCs in longer-lived species occurs in severe hypoxic conditions (Fig. 8.1).

As seen in Fig. 8.1, there is a significant negative correlation between the oviduct Po_2 and species-specific MLS ($r = -0.783$; $p < 10^{-6}$). In fact, HHE at the blastocyte stages of ESCs development should be even stronger than in the oviduct because ESCs of the inner cell mass (ICM) are piled up in several layers. It means that cells in the central localization of the pile can have a lower possibility of O_2/CO_2 exchange.

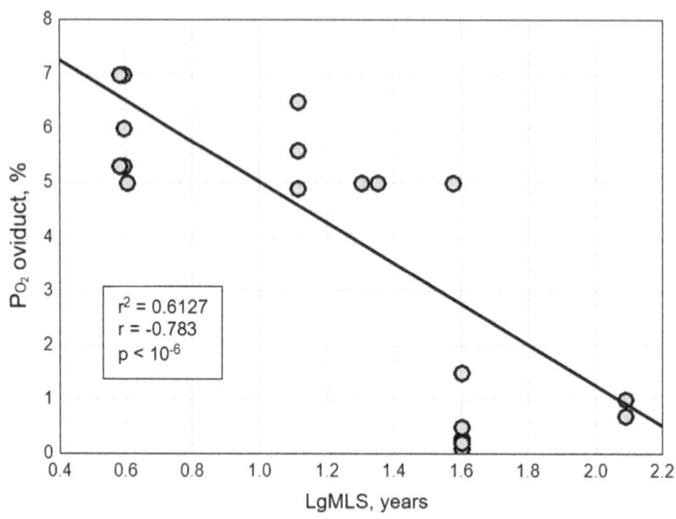

Fig. 8.1 Correlation of Po_2 in the oviduct with MSL in mammals. The Po_2 data were collected from PubMed and the MLS records were taken from the AnAge database (https://genomics.senescence.info/species/index.html; Tacutu et al. 2018)

Additionally, ICM is isolated by protective walls of the blastocyst trophectoderm and primitive endoderm which may consume a significant part of available O_2.

Why is HHE so important for this initial period of embryogenesis, i.e., before ESCs start to differentiate? Such microenvironment is associated with twofold benefits for ESCs. First, HHE decreases the rate of oxidative phosphorylation (OXPHOS) and associated ROS production. It is particularly important in view of an increased sensitivity to death stimuli and completely de-differentiated status of ESCs. Second, the HHE-induced suppression of OXPHOS activates anaerobic metabolism which is necessary for supplying the substrates for the anabolic processes in actively dividing cells, ESCs included (Vander Heiden et al. 2009; Krisher and Prather 2012). Similar metabolic patterns were observed in various models of induced pluripotency which usually started by OXPHOS inhibition and activation of glycolysis (Folmes et al. 2011; Bukowiecki et al. 2014). Remarkably, even at high O_2 levels, ESCs continue to utilize primarily anaerobic metabolism (Powers et al. 2008). The severe HHE in preimplanted blastocysts is required for prevention of ESC differentiation, thus allowing a highly efficient *'last-minute' DNA repair*—a crucial rejuvenative event before ESCs start to differentiate (Muradian et al. 2019). Otherwise, any errors in ESCs will be multiplied in their progeny. Remarkably, most recently, the evidence has been obtained, indicating that the embryo at this 'last-minute' correction may correspond to the period shortly before the minimal biological age (Kerepesi et al. 2021; Kerepesi and Gladyshev 2023).

Could the O_2 and CO_2 levels at the primordial times be optimal for ESC metabolism? Indeed, the oviduct HHE in long-lived mammals resembles the low levels of atmospheric Po_2 and the high ones of Pco_2 at the time of multicellular organisms' origin. Data received from the mid-proterozoic sediments from North America, China, and Australia evidences that atmospheric O_2 was around 0.1% of the present level (Planavsky et al. 2014).

8.3 ESC Growth Rate and Mammalian Longevity

It is widely known that mammalian MLS strongly correlates with the periods of gestation, sexual maturation and growth. Our preliminary analysis shows that a positive correlation also exists between MLS and the ESC development (from zygote to the blastocyst implantation) (Fig. 8.2). Since the final number of ESCs in ICM of preimplanted blastocyte is almost equal across mammalian species, the duration of ESC development in oviduct reflects the ESC growth rate.

The slower rate of ESC growth could be associated with a more severe HHE in the oviduct of longer-lived species, thus causing slow-down of metabolism and decreasing the development rate (Figs. 8.1 and 8.2).

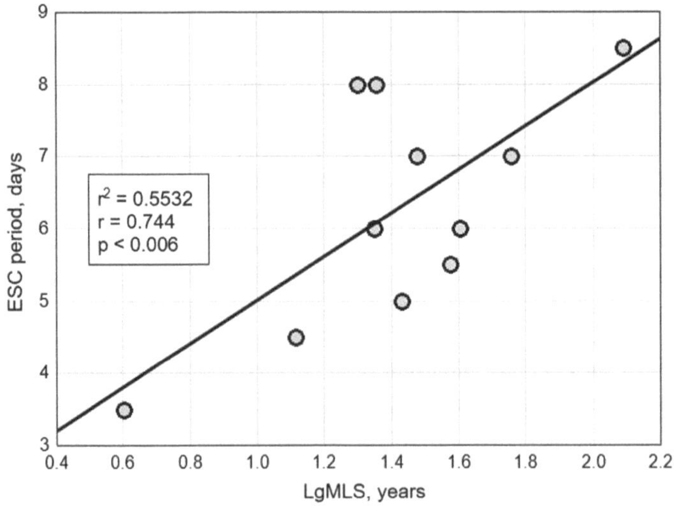

Fig. 8.2 Correlation between the duration of ESC development in oviduct and MLS in mammals. MLS records were taken from AnAge database: https://genomics.senescence.info/species/index.html; Tacutu et al. (2018)

8.4 One ICM Cell—One Type of a Differentiated Cell

It is believed that a typical adult human body of 70 kg includes about 30 trillion cells of 230 types in 79 organs (Sender et al. 2016). Another important, yet poorly studied issue concerns the descendants of these cells. In most explored species in this regard (mice, rhesus monkeys and humans), the period after fertilization to blastocyte implantation is 3.5, 7 and 8.5 days, respectively. During these periods, mouse ESCs divide each 7 h, ESCs of rhesus monkey divide each 15 h and the human ESCs divide each 16 h (Fluckiger et al. 2006; Liu et al. 2019). After subtraction of the log periods for the replication initiation, the total number of cell cycles corresponds to ca. 9 10 doublings. It should be about 600–700 cells of which around one third belongs to ICM. So, despite the different metabolic rate, duration of ESC development and the cell cycle time, the final number of ICM should be about 200–230 cells. We assume that the coincidence between the numbers of ESCs in ICM and cell types in mammalian body is not by chance. Each of these cells is responsible for the arrangement of its specific cell type, according to the principle '***one ESC in ICM—one differentiated cell type in the body***'. For humans, it means that each ESC in ICM gives rise to ca. 150 billion of differentiated cells and, therefore, the cost of any unrepaired error in ESC might have far-reaching consequences.

8.5 Concluding Remarks

Embryonic stem cells (ESCs) represent a small group of elite immortal cells. ESCs are a 'gold standard' for comparison and understanding the basic mechanisms of pluripotency and for the development of new biotechnologies. ESCs display paradoxical combination of superior protection against damaging factors with hypersensitivity to apoptosis. The latter is an additional line of protection of the ESC progeny. The hypoxic-hypercapnic environment is crucial for preserving the ESC features. ESCs reside in HHE and rely on glycolysis.

We assume that the hypoxia and hypercapnia reach the lowest and the highest values, respectively, in the preimplanted blastocyst, prior to differentiation of ESCs. This period provides a unique opportunity for the 'last-minute' correction of the DNA damage. Comparison with epigenetic clock data (Kerepesi et al. 2021; Kerepesi and Gladyshev 2023) indicates that after the 'last-minute' correction embryo reaches the minimal biological age.

Ethics All the materials and data described in the book have been approved by Ethical Committee for Animal Experimentation or Helsinki Committee for Research on Humans of the organizations, where corresponding investigations have been done.

References

Bukowiecki R, Adjaye J, Prigione A (2014) Mitochondrial function in pluripotent stem cells and cellular reprogramming. Gerontology 60(2):174–182. https://doi.org/10.1159/000355050. PMID: 24281332

Evans MJ, Kaufman MH (1981) Establishment in culture of pluripotential cells from mouse embryos. Nature 292(5819):154–156. https://doi.org/10.1038/292154a0. PMID: 7242681

Fischer B, Bavister BD (1993) Oxygen tension in the oviduct and uterus of rhesus monkeys, hamsters and rabbits. J Reprod Fertil 99(2):673–679. https://doi.org/10.1530/jrf.0.0990673. PMID: 8107053

Fluckiger AC, Marcy G, Marchand M, Négre D, Cosset FL, Mitalipov S, Wolf D, Savatier P, Dehay C (2006) Cell cycle features of primate embryonic stem cells. Stem Cells 24(3):547–556. https://doi.org/10.1634/stemcells.2005-0194. PMID: 16239321

Folmes CD, Nelson TJ, Terzic A (2011) Energy metabolism in nuclear reprogramming. Biomark Med 5(6):715–729. https://doi.org/10.2217/bmm.11.87. PMID: 22103608

Hanna JH, Saha K, Jaenisch R (2010) Pluripotency and cellular reprogramming: facts, hypotheses, unresolved issues. Cell 143(4):508–525. https://doi.org/10.1016/j.cell.2010.10.008. PMID: 21074044

Horii T, Hatada I (2015) Genome editing using mammalian haploid cells. Int J Mol Sci 16(10):23604–23614. https://doi.org/10.3390/ijms161023604. PMID: 26437403

Kerepesi C, Gladyshev VN (2023) Intersection clock reveals a rejuvenation event during human embryogenesis. Aging Cell 22(10):e13922. https://doi.org/10.1111/acel.13922. PMID: 37786333

Kerepesi C, Zhang B, Lee SG, Trapp A, Gladyshev VN (2021) Epigenetic clocks reveal a rejuvenation event during embryogenesis followed by aging. Sci Adv 7(26):eabg6082. https://doi.org/10.1126/sciadv.abg6082. PMID: 34172448

Krisher RL, Prather RS (2012) A role for the Warburg effect in preimplantation embryo development: metabolic modification to support rapid cell proliferation. Mol Reprod Dev 79(5):311–320. https://doi.org/10.1002/mrd.22037. PMID: 22431437

Liu L, Michowski W, Kolodziejczyk A, Sicinski P (2019) The cell cycle in stem cell proliferation, pluripotency and differentiation. Nat Cell Biol 21(9):1060–1067. https://doi.org/10.1038/s41556-019-0384-4. PMID: 31481793

Maas DH, Storey BT, Mastroianni L Jr (1977) Hydrogen ion and carbon dioxide content of the oviductal fluid of the rhesus monkey (Macaca mulatta). Fertil Steril 28(9):981–985 PMID: 19307

Martin GR (1981) Isolation of a pluripotent cell line from early mouse embryos cultured in medium conditioned by teratocarcinoma stem cells. Proc Natl Acad Sci USA 78(12):7634–7638. https://doi.org/10.1073/pnas.78.12.7634. PMID: 6950406

Muradian K, Fraifeld V (2023) Embryogenesis of longer-lived mammalian species occurs in a more severe hypoxic-hypercapnic environment. AgingBio 1:1–4. https://doi.org/10.59368/agingbio.20230018

Muradian KK, Tolstun DA, Paier AG, Popa-Wagner A, Fraifeld VE (2019) Embryonic stem cells, telomeres and aging. J Ageing Restor Med (JARM) 2(3):115–122. www.scitcentral.com

Ng KYB, Mingels R, Morgan H, Macklon N, Cheong Y (2018) In vivo oxygen, temperature and pH dynamics in the female reproductive tract and their importance in human conception: a systematic review. Hum Reprod Update 24(1):15–34. https://doi.org/10.1093/humupd/dmx028. PMID: 29077897

Pendse S, Vaidya A, Kale V (2022) Clinical applications of pluripotent stem cells and their derivatives: current status and future perspectives. Regen Med 17(9):677–690. https://doi.org/10.2217/rme-2022-0045. PMID: 35703035

Planavsky NJ, Reinhard CT, Wang X, Thomson D, McGoldrick P, Rainbird RH, Johnson T, Fischer WW, Lyons TW (2014) Earth history. Low mid-Proterozoic atmospheric oxygen levels and the delayed rise of animals. Science 346(6209):635–638. https://doi.org/10.1126/science.1258410. PMID: 25359975

Powers DE, Millman JR, Huang RB, Colton CK (2008) Effects of oxygen on mouse embryonic stem cell growth, phenotype retention, and cellular energetics. Biotechnol Bioeng 101(2):241–254. https://doi.org/10.1002/bit.21986

Rudnitsky E, Braiman A, Muradian KK, Wolfson M, Gorbunova V, Turgeman G, Fraifeld VE (2024) Stem cell-derived exosomes as senotherapeutics. Ageing Res Rev 99:102391. doi: https://doi.org/10.1016/j.arr.2024.102391. PMID: 38914266

Sender R, Fuchs S, Milo R (2016) Revised estimates for the number of human and bacteria cells in the body. PLoS Biol 14(8):e1002533. https://doi.org/10.1371/journal.pbio.1002533. PMID: 27541692

Tacutu R, Thornton D, Johnson E, Budovsky A, Barardo D, Craig T, Diana E, Lehmann G, Toren D, Wang J, Fraifeld VE, de Magalhães JP (2018) Human ageing genomic resources: new and updated databases. Nucleic Acids Res 46(D1):D1083–D1090. https://doi.org/10.1093/nar/gkx1042. PMID: 29121237

Thomson JA, Itskovitz-Eldor J, Shapiro SS, Waknitz MA, Swiergiel JJ, Marshall VS, Jones JM (1998) Embryonic stem cell lines derived from human blastocysts. Science 282(5391):1145–1147. https://doi.org/10.1126/science.282.5391.1145. PMID: 9804556

Vander Heiden MG, Cantley LC, Thompson CB (2009) Understanding the Warburg effect: the metabolic requirements of cell proliferation. Science 324(5930):1029–1033. https://doi.org/10.1126/science.1160809. PMID: 19460998

Yamanaka S (2020) Pluripotent stem cell-based cell therapy-promise and challenges. Cell Stem Cell 27(4):523–531. https://doi.org/10.1016/j.stem.2020.09.014. PMID: 33007237

Yu L, Wen H, Liu C, Wang C, Yu H, Zhang K, Han Q, Liu Y, Han Z, Li Z, Liu N (2023) Embryonic stem cell-derived extracellular vesicles rejuvenate senescent cells and antagonize aging in mice. Bioact Mater 29:85–97. https://doi.org/10.1016/j.bioactmat.2023.06.011. PMID: 37449253

Zhang Z, Bao X, Lin CP (2023) Progress and prospects of gene editing in pluripotent stem cells. Biomedicines 11(8):2168. https://doi.org/10.3390/biomedicines11082168. PMID: 37626665

Chapter 9
Muse Cells and VSEL Stem Cells in Longevity and Rejuvenation

Abstract Multilineage-differentiating stress-enduring cells (Muse cells) and very small embryonic-like (VSEL) stem cells are natural pluripotent stem cells (PSCs). They can be a universal remedy for recovery/rejuvenation of any type of cells in metazoans. Muse cells and VSELs are small, rare and self-renewable. They are quiescent but in stress or injury, they migrate into the damaged site, differentiate into the cells compatible with the homing tissue and rejuvenate the cell population. Muse cells could be easier isolated and appeared to be therapeutically more efficient. That is why they may conquer the biomedicine faster than other stem cells. Survival and proliferation of Muse cells depend on the adherence of their environment. Therefore, it seems that Muse cells 'know' the number of cells to be sent to recover the damage. Although Muse composition and function are not fully identified, their repair efficiency was demonstrated in numerous pre-clinical studies, involving the models of acute myocardial infarction, stroke, amyotrophic lateral sclerosis, wound healing of diabetic ulcers, skin regeneration and various pathological conditions of the liver, kidney, and lung. For unknown reasons, exploration of the aging aspects of Muse cells has thus far been ignored. Muse cells make up only 0.03% of the bone marrow mononucleated cells or a few percent of mesenchymal stem cells in various tissues.

What was the reason for the evolutionary 'greed' for a more generous distribution of these universal rescue cells? What could be done to increase population and robustness of these cells which seem to be crucial for longevity and rejuvenation? There is a lot of similarity between VSELs and Muse cells. However, VSELs are known in gerontology because they have been studied in murine aging models. VSELs content decreased over 17-fold in the bone marrow of mice within 2–36 months of age. Population of these cells can be increased by the calorie-restriction, physical exercising, administration of nicotinamide, valproic acid, and other small-molecules. Damaged cells are nearly everywhere in an aging organism. It is important that these rescue systems are situated nearby in virtually every tissue. We believe that the non-declared competition between iPSCs, Muse cells, VSEL stem cells, ESCs, spore-like cells and other stem cells will be in favor for rejuvenation and health span.

K. K. Muradian and V. E. Fraifeld, *Rejuvenation and Longevity*, Healthy Ageing and Longevity 20, https://doi.org/10.1007/978-3-031-64995-0_9

Keywords Pluripotent stem cells · Tissue recovery · Longevity · Muse cells · Age-related pathology · Aging · Very small embryonic-like (VSEL) stem cells · Spore-like cells

To age, or not to age, that is the question.

9.1 Pluripotent Stem Cells

A pessimist believes that poor health is the cause of aging reluctance, while an optimist says that 'aging is a slowdown of rejuvenation.' The pessimist-optimist challenge could be fateful for the professional starting point. It may resolve the dilemma to deal 'against aging or for rejuvenation'. It is not the same thing. Answering the Shakespearean-like replica "To age, or not to age, that is the question", kindly whisper: "Try rejuvenation!"

There are at least several types of natural PSCs ready to rejuvenate any of the over 220 types of cells in the human body. This gives ample optimism for rejuvenation and health span extension. In normal conditions, PSCs are quiescent but in response to injury or stress, they 'wake up', migrate into the causality zone, become tissue-committed, replicate new cells and repair the damaged tissue. Unfortunately, these rescue teams obviously function by the principle of 'adaptive sufficiency' that often shifts into a 'practical insufficiency' and eventually provokes aging.

Given that aging is mainly driven by cellular damage, timely replacement of damaged cells should offer "the opportunity to decelerate, stop, or reverse aging" (López-Otín et al. 2023; Moskalev et al. 2013). It is important that various PSC-based recovery systems are often situated nearby the damage area, in the same tissue. Such a close neighborhood can double or triple recovery effects to a rescue signal. Even a moderate excess of recovery/damage ratio could mitigate cellular transgression through the critical border 'aging → rejuvenation'.

PSCs have long been considered as 'magic bullets' which could revolutionize the whole field of biomedicine, rejuvenation included. The stem cells derived from embryonic, fetal, or adult tissues serve as sources for new rejuvenated cells. The huge investments opened new opportunities in cellular therapy. Although various types of PSCs are easily accessible from the bone marrow and adipose tissue, they are heterogeneous and their composition is not fully identified (Dezawa 2016; Park et al. 2021). As such, usage of PSCs is often limited because of their inherent tumorigenicity, immunogenicity, and heterogeneity (Yamanaka 2020). Despite the latest successes in PSC biology, many problems of stem cell-based biomedicine and especially the aging aspects remain largely stagnant or unexplored (Trounson and McDonald 2015; Fisch et al. 2017; Bhartiya et al. 2016).

9.2 The Five Giant PSCs for Tissue Recovery and Longevity

There are at least five basic types of PSCs appropriate for cell rejuvenation and tissue recovery: embryonic stem cells (ESCs), induced PSCs (iPSCs), Muse cells, VSEL stem cells and spore-like cells. The 2006–2007 were fateful for the discovery of most of these cells. Indeed, in these years, the group of Prof. Yamanaka discovered the mouse and human iPSCs (Takahashi et al. 2006, 2007), the group of Prof. Ratajczak found out the VSEL stem cells (Ratajczak et al. 2007), and the group of Prof. Dezewa identified and characterized the Muse cells, though the results were first published a bit later (Kuroda et al. 2010). The spore-like cells were identified by Prof. Vacanti even earlier, in 2001, yet their possible role in anti-aging was discussed only in 2007 (Vacanti et al. 2001; Shmilovici 2007). Despite significant efforts, the postulated qualities of the spore-like cells have not been confirmed in stringent independent studies (Dulak et al. 2015).

The destiny of various types of PSCs turned out differently. The Nobel Prize Committee awarded Prof. Yamanaka for iPSCs in 2012, and this field received huge material and financial support. ESCs remained the golden standard for PSCs. Muse cells passed thorough examination and were used in numerous pre-clinical trials. VSEL stem cells have been given intensive laboratory study but not enough for the pre-clinical therapy. The spore-like cells have received the least attention. Anyway, all of them continue to be the main sources for establishing new strategies and hopes in rejuvenation and tissue regeneration. ESCs and adult (resident) stem cells (ASCs) are discussed elsewhere in this book (see Chaps. 8, 10 and 11). Here, we briefly consider the role of Muse cells and VSELs stem cells.

9.3 Multilineage-Differentiating Stress-Enduring Stem Cells (Muse Cells)

Sources and basic capacities of Muse cells

Muse cells are ASCs that have accidentally been discovered as a subpopulation of mesenchymal stem cells (MSCs) by Japanese scientist Mari Dezawa at Tohoku University in Sendai and first published in 2010 (Kuroda et al. 2010). Being quiescent in normal conditions, these cells can be activated by cellular stress both in vivo and in vitro, e.g. by exposure to proteolytic enzyme collagenase, serum deprivation, low temperatures, hypoxia, etc. (Heneidi et al. 2013). Muse cells exhibit robust expression of the pluripotency-related genes Oct4, Sox2, Nanog, Rex1, and PAR4, have the self-renewal capacity and can differentiate into other cells spontaneously or under specific inductions (Dezawa 2016). For example, in mice, Muse cells isolated from various stromal compartments including the bone marrow, adipose tissue, ear connective tissue, showed spontaneous commitment to differentiation into the cells of all three germ cell layers (Aprile et al. 2021). In humans and most animal species studied thus far (mouse, rat, rabbit, swine, goat, sheep, dog), Muse cells were isolated from fresh bone marrow or bone marrow stem cells; from fat after liposuction or adipose tissue-derived stem cells, and dermal fibroblasts (Ogura et al. 2014; Tian et al. 2017; Velasco et al. 2023). In average, Muse cells make up 1–10% of cell populations from which they were isolated and only 0.04% of the bone marrow mononucleated cells (Kuroda et al. 2010, 2013; Ogura et al. 2014; Tian et al. 2017; Yamashita et al. 2021).

Muse cells do not exhibit tumorigenic or teratogenic activity and are stress-resistant

Of crucial importance is that, in contrast to the ESCs and iPSCs, Muse cells do not exhibit tumorigenic or teratogenic activity because they have low replicative performance and telomerase activity (Ogura et al. 2014; Tian et al. 2017; Fisch et al. 2017). Compared with other ASCs, Muse cells are characterized by a 770-fold higher expression of the anti-cancer gene CXCL2 and near 40-fold higher expression of SOD2 and aldehyde dehydrogenase genes which are critical in antioxidant, anti-apoptotic, and detoxification defense (Heneidi et al. 2013; Cao et al. 2020). As mentioned above, to be activated from the quiescent state, Muse cells have to be stressed (Heneidi et al. 2013). They are more resistant to physical and chemical insults than other PSCs, can persist for long periods of genotoxic stress and show lower apoptosis and cellular senescence. By unknown mechanism, Muse cells activate expression of ancient genes typical for the response to cellular stress and acute injury in primitive organisms—*Drosophila* and *Caenorhabditis elegans* in metazoans; *Saccharomyces cerevisiae* and *Chlamydomonas* in unicellular eukaryotes, yeast, and green algae, respectively (Heneidi et al. 2013; Fisch et al. 2017).

Muse cells possess robust capacity in sensing and repairing DNA damage through the base and nucleotide excision repair and repairing double-strand breaks by the nonhomologous end joining process (Squillaro et al. 2018; Alessio et al. 2018; Trosko 2018).

Muse cells in treatment of age-related pathology

Along with their high regenerative potential, Muse cells exert anti-inflammatory activity via reducing the secretion of proinflammatory cytokines, interferon-γ and tumor necrosis factor-α, and increasing the secretion of the anti-inflammatory cytokine IL-10 (Gimeno et al. 2017; Alessio et al. 2017). In addition, Muse cells have immunomodulatory capacity mediated by TGFβ1 (Gimeno et al. 2017). They express the matrix metalloproteinase genes MMP-2 and MMP-9, thus exhibiting anti-fibrotic and fibrinolytic effects. Muse cells release several factors which promote angiogenesis and stimulate endogenous progenitors (Oguma et al. 2022). All this makes Muse cells very attractive as a therapeutic modality for treating a variety of pathological conditions, ARDs included.

Another important point is that Muse cells home more efficiently to the injured site than other PSCs. Due to the higher homing, migration efficiency, the rescue side, and a variety of beneficial biological activities, Muse cells has successfully been investigated in many in vivo and in vitro models, including numerous pre-clinical studies (reviewed by Cao et al. 2020; Velasco et al. 2023). The high efficacy of Muse cells was established in the models of acute myocardial infarction, stroke, amyotrophic lateral sclerosis, aortic aneurysm, type 1 diabetes, diabetic skin ulcers, nephropathy (focal segmental glomerulosclerosis), osteochondral defects, atopic dermatitis, liver fibrosis, and ischemia–reperfusion lung injury.

Stroke is a cerebrovascular disease caused by the arterial occlusion or rupture of cerebral blood vessels. It is characterized by a high incidence, disability, and medical expenses, being the second leading cause of death worldwide (Campbell et al. 2019). Muse cells migrate to the damaged sites of the brain, survive in hostile environment and spontaneously differentiate into the brain tissue-compatible cells (Li et al. 2022). In the acute stage of ischemic stroke, Muse cells were successfully mobilized from bone marrow into peripheral blood (Hori et al. 2016) and then from the peripheral blood migrated into the areas of brain damage (Shimamura et al. 2017). Therapeutic interventions that increase endogenous Muse cells or exogenous administration of Muse cells may improve functional outcome after ischemic stroke.

Muse cells appear to be a promising source for treatment of another ARD, amyotrophic lateral sclerosis—a fatal neurodegenerative disease which is characterized by the progressive loss of motor neurons. Muse cells can selectively home to the damaged sites after intravenous injection, exert pleiotropic effects, recover the number of motor neurons in the spinal cord, and alleviate denervation in lower limb muscles (Yamashita et al. 2020).

The therapeutic value of Muse cells in tissue repair and regeneration was also demonstrated in acute myocardial infarction. Muse cells were twice faster mobilized to injured site than in the myocardium of normal subjects. In vivo, these cells spontaneously differentiated into cardiac and vascular lineages, as evidenced by relevant markers, and also stimulate the endogenous cardiomyocyte progenitors (Tanaka et al. 2018; Cao et al. 2020).

Muse cells can be a promising method for skin regeneration. They promote wound healing of diabetic ulcers by differentiating into dermis and vascular endothelial cells in the upper dermis and into other cell types in the lower and middle dermis (Kinoshita et al. 2015). Therapeutic potential of Muse cells was also demonstrated in the liver, kidney, and lung pathologies (Cao et al. 2020; Velasco et al. 2023).

Muse cells and aging

Aging is a progressive accumulation of cellular damage in many, if not all, tissues, while Muse cells are powerful natural pluripotent stem cells which are resident nearby. Aging cells need such a unique remedy. Surprisingly, the aging-related aspects of Muse cells remain virtually unexplored. Examination of the PubMed dataset using the 'Muse cell' and 'aging' as the key words did not find any references. Only in one study on a small human sampling, it was found that the amount of Muse cells derived from adipose tissue negatively correlated with donor age ($r^2 = 0.75$, $p < 0.05$; 11 donors aged 4–88 years) (Yamauchi et al. 2017). It means that aging patients obviously lose part of the internal Muse cells, since pluripotency of Muse cells remains at the same levels regardless of the age of donors (Yamauchi et al. 2017). Of note, Muse cells can easily be harvested from the adipose tissue, allowing autologous treatment and avoiding problems of transplant rejection (Ossanna et al. 2023; Quintero Sierra et al. 2023).

Numerous advantages of Muse cells ensure their wide entering the clinical trials and the very possibility that they can be used in various biomedical areas much faster than any other types of stem cells (Young 2018).

9.4 Very Small Embryonic-Like (VSEL) Stem Cells

According to intriguing data by Ratajczak et al. (2007, 2011, 2019) adult tissues contain very small embryonic-like (VSEL) pluripotent stem cells as a backup for the tissue-committed stem cells. Most likely, these quiescent cells are remnants of primordial germ cells (PGCs), can be involved in tissue repair in normal conditions, or mobilized upon a tissue injury to support tissue repair.

VSEL stem cells have very small size (3–5 μm in mice and 5–7 μm in humans), high nuclear/cytoplasm ratio and unique pattern of genomic imprinting due to hypo- and hypermethylation of selected genes (Ratajczak et al. 2007, 2019). VSEL stem cells could be isolated from several adult animal organs and tissues including brain, liver, skeletal muscles, heart, kidney, human cord blood and peripheral blood (Shin et al. 2010; Ratajczak et al. 2011). They express the nuclear Oct4, Nanog, Rex1, and stage-specific antigen (SSEA) markers, as well as Stella, Fragilis, and several markers of PGCs and showed the ability to differentiate into cellular lineages of all three germ layers (Shin et al. 2010; Ratajczak et al. 2019).

The regenerative potential of VSEL stem cells sharply decreases in aging. A pool of VSEL stem cells in the bone marrow of C57BL/6 mice gradually decreases from $0.052 \pm 0.018\%$ to $0.003 \pm 0.002\%$ between 2 and36 months of age (Ratajczak et al. 2008a, b, 2012). The amount of these cells was higher in the long-lived C57BL/6 mice, compared to the short-lived DBA/2J mice (Kucia et al. 2006). In aging, the number of VSEL stem cells in adult organs decreased in parallel with the decline in their in vitro ability to form primitive stem cells (Ratajczak et al. 2008a, b, 2012). Examination of human heart tissue, collected from healthy and ischemic heart subjects, showed a presence of VSEL stem cells in adult hearts and a decrease in their number in aging (El-Helw et al. 2020).

The accumulated data evidence that the pool of VSEL stem cells can be increased by the longevity-promoting interventions including calorie-restriction, physical exercising or administration of nicotinamide, valproic acid, and small-molecule UM177. On the other hand, depletion of VSEL stem cells through an insulin/insulin-like growth factor signaling induces premature aging (Ratajczak et al. 2017a, b; Lahlil et al. 2018; Heo et al. 2017).

9.5 Conclusion Remarks

Muse cells and VSEL stem cells are one of the giants of ASCs and have much in common: (i) they were discovered in 2006–2007; (ii) they are relatively abundant in the bone marrow and adipose tissue; (iii) they are small, self-renewal and undifferentiated; (iv) upon injury or stress, they can migrate into the rescue site, spontaneously differentiate into the cells compatible with the homing tissue and recover the damaged tissue. VSEL stem cells presumably have a role in determination of longevity. Comparison of murine strains differing in their lifespan showed that the longer-lived strains have a more abundant pool of VSEL stem cells. Moreover, the pool of VSEL stem cells can be increased by longevity-promoting interventions, while depletion of these cells leads to premature aging. Another type of ASCs, Muse cells, showed the lowest tumorigenicity, the highest stress resistance, and robust DNA repair, as compared to other PSCs. Although the main mechanisms remain mostly unknown, Muse cells are more efficient in tissue regeneration than other PSCs, and thus successfully participated in numerous pre-clinical trials. Despite their unique features, almost nothing is known of what happens with Muse cells in aging. Damaged cells are virtually in all aging tissues, and they are needed in recovery systems like Muse and VSEL stem cells. Why has evolution prevented us from keeping the higher reserves of these cells, so important for tissue regeneration? Whatever the case, stimulation of Muse and VSEL stem cells production and self-protection could be a perspective pathway for rejuvenation and radical health span extension.

Ethics All the materials and data described in the book have been approved by Ethical Committee for Animal Experimentation or Helsinki Committee for Research on Humans of the organizations, where corresponding investigations have been done.

References

Alessio N, Özcan S, Tatsumi K, Murat A, Peluso G, Dezawa M, Galderisi U (2017) The secretome of MUSE cells contains factors that may play a role in regulation of stemness, apoptosis and immunomodulation. Cell Cycle 16(1):33–44. https://doi.org/10.1080/15384101.2016.1211215. PMID: 27463232

Alessio N, Squillaro T, Özcan S, Di Bernardo G, Venditti M, Melone M, Peluso G, Galderisi U (2018) Stress and stem cells: adult muse cells tolerate extensive genotoxic stimuli better than mesenchymal stromal cells. Oncotarget 9(27):19328–19341. https://doi.org/10.18632/oncotarget.25039. PMID: 29721206

Aprile D, Alessio N, Demirsoy IH, Squillaro T, Peluso G, Di Bernardo G, Galderisi U (2021) MUSE stem cells can be isolated from stromal compartment of mouse bone marrow, adipose tissue, and ear connective tissue: a comparative study of their in vitro properties. Cells 10(4):761. https://doi.org/10.3390/cells10040761. PMID: 33808472

Bhartiya D, Shaikh A, Anand S, Patel H, Kapoor S, Sriraman K, Parte S, Unni S (2016) Endogenous, very small embryonic-like stem cells: critical review, therapeutic potential and a look ahead. Hum Reprod Update 23(1):41–76. https://doi.org/10.1093/humupd/dmw030. PMID: 27614362

Campbell BCV, De Silva DA, Macleod MR, Coutts SB, Schwamm LH, Davis SM, Donnan GA (2019) Ischaemic stroke. Nat Rev Dis Primers 5(1):70. https://doi.org/10.1038/s41572-019-0118-8. PMID: 31601801

Cao J, Yang Z, Xiao R, Pan B (2020) Regenerative potential of pluripotent nontumorgenetic stem cells: multilineage differentiating stress enduring cells (Muse cells). Regen Ther 15:92–96. https://doi.org/10.1016/j.reth.2020.04.011. PMID: 33426206

Dezawa M (2016) Muse cells provide the pluripotency of mesenchymal stem cells: direct contribution of muse cells to tissue regeneration. Cell Transplant 25(5):849–861. https://doi.org/10.3727/096368916X690881. PMID: 26884346

Dulak J, Szade K, Szade A, Nowak W, Józkowicz A (2015) Adult stem cells: hopes and hypes of regenerative medicine. Acta Biochim Pol 62(3):329–337. https://doi.org/10.18388/abp.2015_1023. PMID: 26200199

El-Helw M, Chelvarajan L, Abo-Aly M, Soliman M, Milburn G, Conger AL, Campbell K, Ratajczak MZ, Abdel-Latif A (2020) Identification of human very small embryonic like stem cells (VSELS) in human heart tissue among young and old individuals. Stem Cell Rev Rep 16(1):181–185. https://doi.org/10.1007/s12015-019-09923-1. PMID: 31758373

Fisch SC, Gimeno ML, Phan JD, Simerman AA, Dumesic DA, Perone MJ, Chazenbalk GD (2017) Pluripotent nontumorigenic multilineage differentiating stress enduring cells (muse cells): a seven-year retrospective. Stem Cell Res Ther 8(1):227. https://doi.org/10.1186/s13287-017-0674-3. PMID: 29041955

Gimeno ML, Fuertes F, Barcala Tabarrozzi AE, Attorressi AI, Cucchiani R, Corrales L, Oliveira TC, Sogayar MC, Labriola L, Dewey RA, Perone MJ (2017) Pluripotent nontumorigenic adipose tissue-derived muse cells have immunomodulatory capacity mediated by transforming growth factor-β1. Stem Cells Transl Med 6(1):161–173. https://doi.org/10.5966/sctm.2016-0014. PMID: 28170177

Heneidi S, Simerman AA, Keller E, Singh P, Li X, Dumesic DA, Chazenbalk G (2013) Awakened by cellular stress: isolation and characterization of a novel population of pluripotent stem cells derived from human adipose tissue. PLoS ONE 8(6):e64752. https://doi.org/10.1371/journal.pone.0064752. PMID: 23755141

Heo J, Lim J, Lee S, Jeong J, Kang H, Kim Y, Kang JW, Yu HY, Jeong EM, Kim K, Kucia M, Waigel SJ, Zacharias W, Chen Y, Kim IG, Ratajczak MZ, Shin DM (2017) Sirt1 regulates DNA methylation and differentiation potential of embryonic stem cells by antagonizing Dnmt3l. Cell Rep 18(8):1930–1945. https://doi.org/10.1016/j.celrep.2017.01.074. PMID: 28228259

Hori E, Hayakawa Y, Hayashi T, Hori S, Okamoto S, Shibata T, Kubo M, Horie Y, Sasahara M, Kuroda S (2016) Mobilization of Pluripotent Multilineage-Differentiating Stress-Enduring Cells in Ischemic Stroke. J Stroke Cerebrovasc Dis 25(6):1473–1481. https://doi.org/10.1016/j.jstrokecerebrovasdis.2015.12.033. PMID: 27019988

Kinoshita K, Kuno S, Ishimine H, Aoi N, Mineda K, Kato H, Doi K, Kanayama K, Feng J, Mashiko T, Kurisaki A, Yoshimura K (2015) Therapeutic potential of adipose-derived SSEA-3-positive muse cells for treating diabetic skin ulcers. Stem Cells Transl Med 4(2):146–155. https://doi.org/10.5966/sctm.2014-0181. PMID: 25561682

Kucia M, Reca R, Campbell FR, Zuba-Surma E, Majka M, Ratajczak J, Ratajczak MZ (2006) A population of very small embryonic-like (VSEL) CXCR4(+)SSEA-1(+)Oct-4+ stem cells identified in adult bone marrow. Leukemia 20(5):857–858. https://doi.org/10.1038/sj.leu.2404171. PMID: 16498386

Kuroda Y, Kitada M, Wakao S, Nishikawa K, Tanimura Y, Makinoshima H, Goda M, Akashi H, Inutsuka A, Niwa A, Shigemoto T, Nabeshima Y, Nakahata T, Nabeshima Y, Fujiyoshi Y, Dezawa M (2010) Unique multipotent cells in adult human mesenchymal cell populations. Proc Natl Acad Sci USA 107(19):8639–8643. https://doi.org/10.1073/pnas.0911647107. PMID: 20421459

Kuroda Y, Wakao S, Kitada M, Murakami T, Nojima M, Dezawa M (2013) Isolation, culture and evaluation of multilineage-differentiating stress-enduring (Muse) cells. Nat Protoc 8(7):1391–1415. https://doi.org/10.1038/nprot.2013.076. PMID: 23787896

Lahlil R, Scrofani M, Barbet R, Tancredi C, Aries A, Hénon P (2018) VSELs maintain their pluripotency and competence to differentiate after enhanced ex vivo expansion. Stem Cell Rev Rep 14(4):510–524. https://doi.org/10.1007/s12015-018-9821-1. PMID: 29736843

Li H, Wei J, Liu X, Zhang P, Lin J (2022) Muse cells: ushering in a new era of stem cell-based therapy for stroke. Stem Cell Res Ther 13(1):421. https://doi.org/10.1186/s13287-022-03126-1. PMID: 35986359

López-Otín C, Blasco MA, Partridge L, Serrano M, Kroemer G (2023) Hallmarks of aging: an expanding universe. Cell 186(2):243–278. https://doi.org/10.1016/j.cell.2022.11.001. PMID: 36599349

Moskalev AA, Shaposhnikov MV, Plyusnina EN, Zhavoronkov A, Budovsky A, Yanai H, Fraifeld VE (2013) The role of DNA damage and repair in aging through the prism of Koch-like criteria. Ageing Res Rev 12(2):661–684. https://doi.org/10.1016/j.arr.2012.02.001. PMID: 22353384

Ogura F, Wakao S, Kuroda Y, Tsuchiyama K, Bagheri M, Heneidi S, Chazenbalk G, Aiba S, Dezawa M (2014) Human adipose tissue possesses a unique population of pluripotent stem cells with nontumorigenic and low telomerase activities: potential implications in regenerative medicine. Stem Cells Dev 23(7):717–728. https://doi.org/10.1089/scd.2013.0473. PMID: 24256547

Oguma Y, Kuroda Y, Wakao S, Kushida Y, Dezawa M (2022) Single-cell RNA sequencing reveals different signatures of mesenchymal stromal cell pluripotent-like and multipotent populations. iScience 25(11):105395. https://doi.org/10.1016/j.isci.2022.105395. PMID: 36339265

Ossanna R, Veronese S, Quintero Sierra LA, Conti A, Conti G, Sbarbati A (2023) Multilineage-differentiating stress-enduring cells (muse cells): an easily accessible, pluripotent stem cell niche with unique and powerful properties for multiple regenerative medicine applications. Biomedicines 11(6):1587. https://doi.org/10.3390/biomedicines11061587. PMID: 37371682

Park YJ, Farooq J, Cho J, Cozene B, Gonzales-Portillo B, Sadanandan N, Saft M, Lee JY, Borlongan CV (2021) A museum of stem cells points to muse cells as robust transplantable cells for stroke: review. Adv Exp Med Biol 1312:165–177. https://doi.org/10.1007/5584_2020_596. PMID: 33159307

Quintero Sierra LA, Biswas R, Conti A, Busato A, Ossanna R, Zingaretti N, Parodi PC, Conti G, Riccio M, Sbarbati A, De Francesco F (2023) Highly pluripotent adipose-derived stem cell-enriched nanofat: a novel translational system in stem cell therapy. Cell Transplant 32:9636897231175968. https://doi.org/10.1177/09636897231175968. PMID: 37243545

Ratajczak MZ, Zuba-Surma EK, Machalinski B, Kucia M (2007) Bone-marrow-derived stem cells–our key to longevity? J Appl Genet 48(4):307–319. https://doi.org/10.1007/BF03195227. PMID: 17998587

Ratajczak MZ, Zuba-Surma EK, Shin DM, Ratajczak J, Kucia M (2008a) Very small embryonic-like (VSEL) stem cells in adult organs and their potential role in rejuvenation of tissues and longevity. Exp Gerontol 43(11):1009–1017. https://doi.org/10.1016/j.exger.2008.06.002. PMID: 18601995

Ratajczak MZ, Zuba-Surma EK, Wysoczynski M, Ratajczak J, Kucia M (2008b) Very small embryonic-like stem cells: characterization, developmental origin, and biological significance. Exp Hematol 36(6):742–751. https://doi.org/10.1016/j.exphem.2008.03.010. PMID: 18474305

Ratajczak MZ, Liu R, Ratajczak J, Kucia M, Shin DM (2011) The role of pluripotent embryonic-like stem cells residing in adult tissues in regeneration and longevity. Differentiation 81(3):153–161. https://doi.org/10.1016/j.diff.2011.01.006. PMID: 21339038

Ratajczak MZ, Shin DM, Liu R, Mierzejewska K, Ratajczak J, Kucia M, Zuba-Surma EK (2012) Very small embryonic/epiblast-like stem cells (VSELs) and their potential role in aging and organ rejuvenation–an update and comparison to other primitive small stem cells isolated from adult tissues. Aging (Albany NY) 4(4):235–246. https://doi.org/10.18632/aging.100449. PMID: 22498452

Ratajczak MZ, Bartke A, Darzynkiewicz Z (2017a) Prolonged growth hormone/insulin/insulin-like growth factor nutrient response signaling pathway as a silent killer of stem cells and a culprit in aging. Stem Cell Rev Rep 13(4):443–453. https://doi.org/10.1007/s12015-017-9728-2

Ratajczak MZ, Ratajczak J, Suszynska M, Miller DM, Kucia M, Shin DM (2017b) A novel view of the adult stem cell compartment from the perspective of a quiescent population of very small embryonic-like stem cells. Circ Res 120(1):166–178. https://doi.org/10.1161/CIRCRESAHA.116.309362. PMID: 28057792

Ratajczak MZ, Ratajczak J, Kucia M (2019) Very small embryonic-like stem cells (VSELs). Circ Res 124(2):208–210. https://doi.org/10.1161/CIRCRESAHA.118.314287. PMID: 30653438

Shimamura N, Kakuta K, Wang L, Naraoka M, Uchida H, Wakao S, Dezawa M, Ohkuma H (2017) Neuro-regeneration therapy using human Muse cells is highly effective in a mouse intracerebral hemorrhage model. Exp Brain Res 235(2):565–572. https://doi.org/10.1007/s00221-016-4818-y. PMID: 27817105

Shin DM, Liu R, Klich I, Ratajczak J, Kucia M, Ratajczak MZ (2010) Molecular characterization of isolated from murine adult tissues very small embryonic/epiblast like stem cells (VSELs). Mol Cells 29(6):533–538. https://doi.org/10.1007/s10059-010-0081-4. PMID: 20526817

Shmilovici A (2007) Mammalian spore-like cells–a reservoir of spare parts for old-age? Med Hypotheses 68(4):767–769. https://doi.org/10.1016/j.mehy.2006.08.044. PMID: 17056185

Squillaro T, Alessio N, Di Bernardo G, Özcan S, Peluso G, Galderisi U (2018) Stem cells and DNA repair capacity: muse stem cells are among the best performers. Adv Exp Med Biol 1103:103–113. https://doi.org/10.1007/978-4-431-56847-6_5. PMID: 30484225

Tanaka T, Nishigaki K, Minatoguchi S, Nawa T, Yamada Y, Kanamori H, Mikami A, Ushikoshi H, Kawasaki M, Dezawa M, Minatoguchi S (2018) Mobilized muse cells after acute myocardial infarction predict cardiac function and remodeling in the chronic phase. Circ J 82(2):561–571. https://doi.org/10.1253/circj.CJ-17-0552. PMID: 28931784

Takahashi K, Yamanaka S (2006) Induction of pluripotent stem cells from mouse embryonic and adult fibroblast cultures by defined factors. Cell 126(4):663–676. https://doi.org/10.1016/j.cell.2006.07.024. PMID: 16904174

Takahashi K, Tanabe K, Ohnuki M, Narita M, Ichisaka T, Tomoda K, Yamanaka S (2007) Induction of pluripotent stem cells from adult human fibroblasts by defined factors. Cell 131(5):861–872. https://doi.org/10.1016/j.cell.2007.11.019. PMID: 18035408

Tian T, Zhang RZ, Yang YH, Liu Q, Li D, Pan XR (2017) Muse cells derived from dermal tissues can differentiate into melanocytes. Cell Reprogram 19(2):116–122. https://doi.org/10.1089/cell.2016.0032. PMID: 28170296

Trosko JE (2018) The role of the mitochondria in the evolution of stem cells, including muse stem cells and their biology. Adv Exp Med Biol 1103:131–152. https://doi.org/10.1007/978-4-431-56847-6_7. PMID: 30484227

Trounson A, McDonald C (2015) Stem cell therapies in clinical trials: progress and challenges. Cell Stem Cell 17(1):11–22. https://doi.org/10.1016/j.stem.2015.06.007. PMID: 26140604

Vacanti MP, Roy A, Cortiella J, Bonassar L, Vacanti CA (2001) Identification and initial characterization of spore-like cells in adult mammals. J Cell Biochem 80(3):455–460. PMID: 11135375

Velasco MG, Satué K, Chicharro D, Martins E, Torres-Torrillas M, Peláez P, Miguel-Pastor L, Del Romero A, Damiá E, Cuervo B, Carrillo JM, Cugat R, Sopena JJ, Rubio M (2023) Multilineage-differentiating stress-enduring cells (muse cells): the future of human and veterinary regenerative medicine. Biomedicines 11(2):636. https://doi.org/10.3390/biomedicines11020636. PMID: 36831171

Yamanaka S (2020) Pluripotent stem cell-based cell therapy-promise and challenges. Cell Stem Cell 27(4):523–531. https://doi.org/10.1016/j.stem.2020.09.014. PMID: 33007237

Yamashita T, Kushida Y, Wakao S, Tadokoro K, Nomura E, Omote Y, Takemoto M, Hishikawa N, Ohta Y, Dezawa M, Abe K (2020) Therapeutic benefit of Muse cells in a mouse model of amyotrophic lateral sclerosis. Sci Rep 10(1):17102. https://doi.org/10.1038/s41598-020-74216-4. PMID: 33051552

Yamashita T, Kushida Y, Abe K, Dezawa M (2021) Non-tumorigenic pluripotent reparative muse cells provide a new therapeutic approach for neurologic diseases. Cells 10(4):961. https://doi.org/10.3390/cells10040961. PMID: 33924240

Yamauchi T, Yamasaki K, Tsuchiyama K, Koike S, Aiba S (2017) A quantitative analysis of multilineage-differentiating stress-enduring (Muse) cells in human adipose tissue and efficacy of melanocytes induction. J Dermatol Sci 86(3):198–205. https://doi.org/10.1016/j.jdermsci.2017.03.001. PMID: 28292562

Young W (2018) Future of muse cells. Adv Exp Med Biol 1103:309–315. https://doi.org/10.1007/978-4-431-56847-6_18. PMID: 30484238

Chapter 10
The Paradox of the Adult Stem Cells: Relevance to Aging and Rejuvenation

Abstract Adult stem cells (ASCs) are mostly quiescent but can quickly react and generate differentiated cells upon a tissue injury or disease. ASCs are indispensable in both rejuvenation and aging. On the one hand, ASCs are tireless 'workhorses' of cellular rejuvenation by replacing worn-out cells with brand-new differentiated cells. On the other hand, ASCs support a tissue cell balance by generating various types of differentiated cells but, paradoxically, cannot save their own tiny population from the age-associated exhaustion. The roots of this ASCs paradox originate from the ambiguity of asymmetric mitosis. In every round of replication, an ASC can generate a group of differentiated cells and a new stem cell but is fundamentally unable to restore its own sporadic losses. Recovery of ASC pool is possible via combination of symmetric and asymmetric replications, as it apparently occurs in development or cancer. We suggest that ASCs in old metazoans either cannot simultaneously afford both modes of replication or, more probably, regulatory systems of higher priority, including antitumor defense, prevent the combined modes of symmetric and asymmetric replications. We posit that intermittent dedifferentiation, spontaneous or artificial, of the only partially differentiated progenitor cells back to multipotent ASCs may induce partial rejuvenation and mitigate negative effects of the ASCs' depletion in aging.

Keywords Adult stem cells (ASCs) · ASCs quiescence · ASCs replication · Metabolism · Mitochondria · Age-related changes · Cell reprogramming · Cell dedifferentiation

10.1 Adult Stem Cells (ASCs): Brief Overview

In an adult multicellular organism, millions and billions of cells are lost daily and miraculously recovered. Adult stem cells (ASCs), also known as somatic or resident stem cells, are the main 'miracle makers.' They are heavily involved in anti-aging and rejuvenation, 'day and night' replacing the worn-out old cells by newly generated differentiated cells. Yet, *successfully controlling the balance between the various*

K. K. Muradian and V. E. Fraifeld, *Rejuvenation and Longevity*, Healthy Ageing and Longevity 20, https://doi.org/10.1007/978-3-031-64995-0_10

types of differentiated cells within a tissue, ASCs are principally unable to save their own tiny population from exhaustion in aging, thus essentially reducing a regenerative potential and eventually organism viability (Moskalev et al. 2013; Rando and Jones 2021; Mi et al. 2022). As such, this is the *main paradox of ASCs.*

ASCs are ubiquitous, rare, small, low-metabolic, lineage-restricted, quiescent, long-lived, multipotent, or progenitors (Rinkevich et al. 2022). They are well 'hidden' in tissue and protected by their microenvironment, known as 'stem cell niches'. Molecular mechanisms that control the ASCs' self-renewal and progeny differentiation are largely related to the surrounding niche (Cheung and Rando 2013; Naik et al. 2018).

10.2 ASCs Quiescence

ASCs tend to be in the G0 quiescence—a prolonged but reversible state of cell cycle arrest. Quiescence has long been viewed as a passive anti-aging protection due to the dormancy and low metabolic activity. Recent research, however, suggests that quiescence is an actively maintained state. Defects in quiescence can lead to accelerated exhaustion of ASCs, compromised tissue renewal and eventually aging (van Velthoven and Rando 2019; Marescal and Cheeseman 2020). Anyway, quiescence is a major self-protection for ASCs pool in the sense that this state prevents ASCs from differentiation (Rumman et al. 2015).

Dormant ASCs can, however, quickly react to tissue injury or other environmental signals by re-entering the cell cycle and inducing cell differentiation (van Velthoven and Rando 2019; Urbán and Cheung 2021). The reversible 'jumping' between dormancy and active proliferation, essential for ASCs, could be inherited from prokaryotes and unicellular eukaryotes which used this 'jumping' far before the metazoans' origin (see Chap. 4). Also, it could indicate that the quick transposition of ASCs between dormant and active states could be a primordial and evolutionarily well-shaped process.

The content of quiescent ASCs varies across species and tissues. It is generally higher in the post-replicative tissues, e.g., skeletal muscle or brain, compared with replicatively active tissues such as the skin or the gut (Post and Clevers 2019; Urbán and Cheung 2021). Little is known about species-specific characteristics of ASCs and their impact on longevity. There are, however, examples indicating that a high content and functionality of stem cells could ensure superior regenerative ability and extremely long life span (e.g., sponges, corals and presumably hydra; see Chap. 5). ASCs are well 'hidden' and thus may keep many secrets of aging and rejuvenation (☺).

10.3 ASCs, Metabolism, and Mitochondria

Although ASCs are mostly quiescent, their metabolic profiles resemble those of proliferating cells rather than of nondividing cells (Coller 2019). This could presumably be because ASCs are intrinsically ready to activate the proliferation machinery within a minimal lag period in response to the injury or other alarm signals. Nevertheless, ASCs exhibit reduced OXPHOS and mitochondria content and rely heavily on glycolysis for energy generation, like most other stem cells do (Vander Heiden et al. 2009; Muradian et al. 2019). Ironically, mitophagy in ASCs is often focused on elimination of not the defective but rather active mitochondria. Anyway, mitochondria are important regulators of ASCs, and mitochondrial dysfunction may compromise ASCs (Ahlqvist et al. 2015). Renewal and metabolism of ASCs could be modified by various intrinsic and environmental factors, among them are hypoxia and hypoxia-inducible factors. They play an important role in self-renewal of ASCs in a variety of tissues (Suda et al. 2011; Ito and Ito 2016).

10.4 ASCs Replication

Asymmetric replication of an ASC gives rise to a new ASC which remains in the ASC pool and a self-renewing progenitor cell. The progenitor daughter can initiate tightly controlled series of symmetric mitosis yielding a group of 4–8 differentiated cells of one or more types (Sánchez-Taltavull 2016; Chen et al. 2016; Rinkevich et al. 2022). It is obviously one of the most efficient ways for multiplication of differentiated cells.

Summarizing, symmetric mitosis can increase the number of ASCs but does not directly affect differentiated cells, while asymmetric mitosis can give rise to new differentiated cells but cannot increase the number of ASCs (Urbán and Cheung 2021). Why has evolution from two modes of mitosis, symmetric and asymmetric, which compensate one another, left predominantly asymmetric one in advanced age? Is it a way by which evolution eliminates the old individuals from population? Could the correction of this obvious 'delusion' of nature ensure cardinal life span extension and rejuvenation?

10.5 Age-Related Changes in ASCs

Every new mammalian life starts from a series of symmetric divisions of the embryonic toti- and pluripotent stem cells. In the subsequent fetal and postnatal development, simultaneous symmetric and asymmetric mitoses can generate new differentiated cells in parallel with maintaining the stem cells' pool. The increasing needs in differentiated cells may result in strengthening the asymmetric replication up to its almost complete dominance in ASCs. The pool size and activity of ASCs can, however, decline in harmful environment, disease, or aging (Schultz and Sinclair 2016; Sameri et al. 2020).

The two main responsibilities of ASCs associated with maintenance of their own small population and generation of tissue-specific differentiated cells, tend to be failed in aging. Notably, ASC pools do not undergo just moderate age-related changes (as most variables do) but rather become 'exhausted'. How do cells with capacity of unlimited replication and mandates for maintaining the pools of differentiated cells get so obsolete that cannot ensure stability of their own tiny cellular community (often within 0.01–0.0001%)? Whatever the explanation, *age-related depletion of ASCs is typical for all or for the vast majority of tissues and species.*

10.6 Are ASCs Potentially Immortal?

All prokaryotes and unicellular eukaryotes remind stem cells in the sense that they can indefinitely generate sister cells in vitro. Although still disputable, ASCs may possess unlimited replicative capacity in culture and hence are considered potentially immortal. They remain young, while progeny cells differentiate, age, and die (Sheldrake 2022). Nevertheless, ASCs' immortality is not absolute and, in particular, depends on mode of partitioning (Colozza et al. 2022).

In symmetric mitosis, the damaged constituents accumulated in the mother stem cell, are equally distributed between the sister stem cells. This mode of division is typical for most immortal prokaryotes (Sheldrake 2022). In asymmetric division of unicellular eukaryotes, a newly emerged and usually smaller daughter cell gets the smaller part of damaged products and rejuvenates at the expense of the larger mother cell. This mode of mitosis is typical for species which propagate by budding, e.g., baker's yeast (Budovsky et al. 2011; see Chap. 2, Fig. 2.1). However, ASCs show that in the world of cells, the mother-daughter relations are much more pragmatic. In contrast to the yeast, a mother ASC is smaller and gives the larger part of accumulated damage to the daughter progenitor cell which differentiates and becomes mortal. In this way, the mother ASC rejuvenates at the expense of the progenitor cell. Thus, ASCs could be inherently immortal. However, due to the unique task-dedication and plasticity they can switch on the mortal mode of replication, if necessary (Sheldrake 2022).

10.7 Mitotic 'Glider'

To expand the cell pools during development, the stem cells have to use both symmetric and asymmetric divisions (Morrison and Kimble 2006), whereas in an advanced age, ASCs mostly use the asymmetric mitosis (Shahriyari and Komarova 2013). We suggest that ASCs in an old metazoan organism either cannot afford simultaneously both symmetric and asymmetric modes of replication, or, more probably, it is prevented by some regulatory systems of higher priority, such as anti-cancer defense.

Asymmetric mode of replication assumes inevitable depletion ASCs because of their incapability to recover the pool. Consequently, it will induce the aging cascade: deficit of ASCs → deficit of differentiated cells → decreased viability → aging → death. One thing is clear: in the case of ASCs, evolution has applied its favored principle of the '*gerontological glider*' which after gaining maximum functionality around the maturity begins to descend before falling to the ground.

How can we get around this principle? There is an intriguing possibility to increase the number of ASCs via intermittent dedifferentiation of the slightly differentiated progenitor cell. This possibility is supported by the fact that among the various cell types, ASCs are the ones which can be easily converted to pluripotency. ASCs are much easier to reprogram, among other things, because they express the pluripotency-inducing transcription factors (Bazley et al. 2015; Sundaravadivelu et al. 2022).

10.8 Concluding Remarks

ASCs are indispensable in cellular rejuvenation because of their key role in replacement of the worn-out cells by the newly replicated and differentiated cells. The apparent paradox is that ASCs are simultaneously involved in two processes virtually opposite by the essence—rejuvenation and aging.

Being dormant in normal conditions, ASCs can quickly become activated and supply the injured tissue(s) with a necessary spectrum of differentiated cells. The repeated switching between the active/dormant states is an evolutionarily primordial invention widely used in prokaryotes and unicellular eukaryotes. Therefore, it should be a well-shaped process in metazoans as well.

The life of most metazoans is initiated by a series of symmetric replication of toti- and pluripotent ESCs, gradually changes into the mixed mode of symmetric/asymmetric mitoses during embryonic and postnatal development and ultimately transfers into predominantly asymmetric partitioning later in life. The attenuation of symmetric replication may explain the inevitability of ASC pool depletion in the advanced age. Moreover, ASCs exhaustion could be a leading cause of declined viability and death of multicellular organisms. Why it happened in evolution remains unknown. However, quasi-immortal species and species growing throughout the whole life evidence that simultaneous mode of symmetric and asymmetric mitosis in adult animals is not only possible but could be associated with healthy longevity.

Ethics All the materials and data described in the book have been approved by Ethical Committee for Animal Experimentation or Helsinki Committee for Research on Humans of the organizations, where corresponding investigations have been done.

References

Ahlqvist KJ, Suomalainen A, Hämäläinen RH (2015) Stem cells, mitochondria and aging. Biochim Biophys Acta 1847(11):1380–1386. https://doi.org/10.1016/j.bbabio.2015.05.014. PMID: 2601437

Bazley FA, Liu CF, Yuan X, Hao H, All AH, De Los AA, Zambidis ET, Gearhart JD, Kerr CL (2015) Direct reprogramming of human primordial germ cells into induced pluripotent stem cells: efficient generation of genetically engineered germ cells. Stem Cells Dev. 24(22):2634–2648. https://doi.org/10.1089/scd.2015.0100. PMID: 26154167

Budovsky A, Fraifeld VE, Aronov S (2011) Linking cell polarity, aging and rejuvenation. Biogerontology 12(2):167–175. https://doi.org/10.1007/s10522-010-9305-4. PMID: 20978937

Chen C, Fingerhut JM, Yamashita YM (2016) The ins(ide) and outs(ide) of asymmetric stem cell division. Curr Opin Cell Biol 43:1–6. https://doi.org/10.1016/j.ceb.2016.06.001. PMID: 27318429

Cheung TH, Rando TA (2013) Molecular regulation of stem cell quiescence. Nat Rev Mol Cell Biol 14(6):329–340. https://doi.org/10.1038/nrm3591. PMID: 23698583

Coller HA (2019) The paradox of metabolism in quiescent stem cells. FEBS Lett 593(20):2817–2839. https://doi.org/10.1002/1873-3468.13608. PMID: 31531979

Colozza G, Park SY, Koo BK (2022) Clone wars: from molecules to cell competition in intestinal stem cell homeostasis and disease. Exp Mol Med 54(9):1367–1378. https://doi.org/10.1038/s12276-022-00854-5. PMID: 36117218

Ito K, Ito K (2016) Metabolism and the Control of Cell Fate Decisions and Stem Cell Renewal. Annu Rev Cell Dev Biol 32:399–409. https://doi.org/10.1146/annurev-cellbio-111315-125134. PMID: 27482603

Marescal O, Cheeseman IM (2020) Cellular mechanisms and regulation of quiescence. Dev Cell 55(3):259–271. https://doi.org/10.1016/j.devcel.2020.09.029. PMID: 33171109

Mi L, Hu J, Li N, Gao J, Huo R, Peng X, Zhang N, Liu Y, Zhao H, Liu R, Zhang L, Xu K (2022) The mechanism of stem cell aging. Stem Cell Rev Rep 18(4):1281–1293. https://doi.org/10.1007/s12015-021-10317-5. PMID: 35000109

Morrison SJ, Kimble J (2006) Asymmetric and symmetric stem-cell divisions in development and cancer. Nature 441(7097):1068–1074. https://doi.org/10.1038/nature04956. PMID: 16810241

Moskalev AA, Shaposhnikov MV, Plyusnina EN, Zhavoronkov A, Budovsky A, Yanai H, Fraifeld VE (2013) The role of DNA damage and repair in aging through the prism of Koch-like criteria. Ageing Res Rev 12(2):661–684. https://doi.org/10.1016/j.arr.2012.02.001. PMID: 22353384

Muradian KK, Tolstun DA, Paier AG, Popa-Wagner A, Fraifeld VE (2019) Embryonic stem cells, telomeres and aging. J Ageing Restor Med (JARM) 2(3):115–122. www.scitcentral.com

Naik S, Larsen SB, Cowley CJ, Fuchs E (2018) Two to tango: dialog between immunity and stem cells in health and disease. Cell 175(4):908–920. https://doi.org/10.1016/j.cell.2018.08.071. PMID: 30388451

Post Y, Clevers H (2019) Defining adult stem cell function at its simplest: the ability to replace lost cells through mitosis. Cell Stem Cell 25(2):174–183. https://doi.org/10.1016/j.stem.2019.07.002. PMID: 31374197

Rando TA, Jones DL (2021) Regeneration, Rejuvenation, and Replacement: Turning Back the Clock on Tissue Aging. Cold Spring Harb Perspect Biol. 13(9):a040907. https://doi.org/10.1101/cshperspect.a040907. PMID: 34187808

Rinkevich B, Ballarin L, Martinez P, Somorjai I, Ben-Hamo O, Borisenko I, Berezikov E, Ereskovsky A, Gazave E, Khnykin D, Manni L, Petukhova O, Rosner A, Röttinger E, Spagnuolo A, Sugni M, Tiozzo S, Hobmayer B (2022) A pan-metazoan concept for adult stem cells: the wobbling Penrose landscape. Biol Rev Camb Philos Soc 97(1):299–325. https://doi.org/10.1111/brv.12801. PMID: 34617397

Rumman M, Dhawan J, Kassem M (2015) Concise review: quiescence in adult stem cells: biological significance and relevance to tissue regeneration. Stem Cells 33(10):2903–2912. https://doi.org/10.1002/stem.2056. PMID: 26075660

Sameri S, Samadi P, Dehghan R, Salem E, Fayazi N, Amini R (2020) Stem cell aging in lifespan and disease: a state-of-the-art review. Curr Stem Cell Res Ther 15(4):362–378. https://doi.org/10.2174/1574888X15666200213105155. PMID: 32053079

Sánchez-Taltavull D (2016) Optimal architecture of differentiation cascades with asymmetric and symmetric stem cell division. J Theor Biol 407:106–117. https://doi.org/10.1016/j.jtbi.2016.07.029. PMID: 27452530

Schultz MB, Sinclair DA (2016) When stem cells grow old: phenotypes and mechanisms of stem cell aging. Development 143(1):3–14. https://doi.org/10.1242/dev.130633. PMID: 26732838

Shahriyari L, Komarova NL (2013) Symmetric vs. asymmetric stem cell divisions: an adaptation against cancer? PLoS One 8(10):e76195. https://doi.org/10.1371/journal.pone.0076195. PMID: 24204602

Sheldrake AR (2022) Cellular senescence, rejuvenation and potential immortality. Proc Biol Sci 289(1970):20212434. https://doi.org/10.1098/rspb.2021.2434. PMID: 35232226

Suda T, Takubo K, Semenza GL (2011) Metabolic regulation of hematopoietic stem cells in the hypoxic niche. Cell Stem Cell 9(4):298–310. https://doi.org/10.1016/j.stem.2011.09.010. PMID: 21982230

Sundaravadivelu PK, Raina K, Thool M, Ray A, Joshi JM, Kaveeshwar V, Sudhagar S, Lenka N, Thummer RP (2022) Tissue-restricted stem cells as starting cell source for efficient generation of pluripotent stem cells: an overview. Adv Exp Med Biol 1376:151–180. https://doi.org/10.1007/5584_2021_660. PMID: 34611861

Urbán N, Cheung TH (2021) Stem cell quiescence: the challenging path to activation. Development 148(3):dev165084. https://doi.org/10.1242/dev.165084. PMID: 33558315

van Velthoven CTJ, Rando TA (2019) Stem cell quiescence: dynamism, restraint, and cellular idling. Cell Stem Cell 24(2):213–225. https://doi.org/10.1016/j.stem.2019.01.001. PMID: 30735649

Vander Heiden MG, Cantley LC, Thompson CB (2009) Understanding the Warburg effect: the metabolic requirements of cell proliferation. Science 324(5930):1029–1033. https://doi.org/10.1126/science.1160809. PMID: 19460998

Chapter 11
Biological Age Can Be Moved in Both Directions

Abstract There is a dynamic equilibrium between aging and rejuvenation factors. In aging organism, aging/rejuvenation ratio (r_{ar}) is > 1, whereas in rejuvenating organism $r_{ar} < 1$. The cell has remarkable potential of immortality, unfortunately, not fully used in metazoans and humans. Stimulation of rejuvenation factors together with a gradual suppression of aging can be a robust strategy for rejuvenation ($r_{ar} \leq 1$). Rejuvenology is a science with growing methodology and models. Biological clocks are one of them, allowing quantitatively assessing the efficiency of rejuvenation factors. There are at least a half-dozen aging clocks, of them, epigenetic clocks are the best known and most reliable. Methylation in selected sequences of DNA, known as epigenetic or biological aging clocks, can be used for estimation of biological age and aging rate. This was shown in both human and animal studies. Epigenetic methylation clock is more sensitive to longevity and mortality, i.e. to biological age, whereas transcriptomic, proteomic and metabolomic clocks show higher sensitivity to disease phenotypes.

Keywords Biological age · Aging/rejuvenation ratio · Biological aging clocks · Cytosine methylation · Epigenome · Decision-making structures · Rejuvenation interventions

11.1 The Plasticity of Biological Age

Recent progress clearly shows that cellular aging is not a one-way street. Indeed, biological clocks indicate that aging rate could be accelerated, delayed, or even reversed. There are numerous examples of rapid transition between aging and rejuvenation at the cellular level (Damri et al. 2009; Simpson et al. 2021; Gill et al. 2022; Chondronasiou et al. 2022; Ji et al. 2023; Puri and Wagner 2023; Yang et al. 2023). Moreover, a growing body of evidence indicates that such transition could occur at organismal level as well. Heterochronic parabiosis (HP), blood plasma exchange, diet and healthy lifestyle, pharmacological treatments, in vivo cell reprogramming with defined transcription factors clearly demonstrated that biological age can be

K. K. Muradian and V. E. Fraifeld, *Rejuvenation and Longevity*, Healthy Ageing and Longevity 20, https://doi.org/10.1007/978-3-031-64995-0_11

decreased (Table 11.1). Of note, stress (e.g., caused by HP, surgery, pregnancy, or severe infections) can increase biological age but it has been restored or even decreased upon recovery (Poganik et al. 2023; Zhang et al. 2023). This response may be age-dependent. For example, according to epigenetic, transcriptomic and metabolomic data, the young partner of HP undergoes accelerated aging but reverses its biological age after detachment of the pair. The old partner of HP becomes epigenetically younger and the decrease in its biological age remains for at least two months after the pair detachment (Table 11.1). The important point is that the older partner lives longer than the control animals of the same age, thus demonstrating that parabiotic rejuvenation has long-lasting effects (Zhang et al. 2023).

11.2 The Aging/Rejuvenation Ratio (r_{ar})

We assume that there is a dynamic equilibrium between factors of aging and rejuvenation. In aging organisms, the ratio of aging/rejuvenation factors (r_{ar}) should be > 1, whereas in rejuvenating organisms $r_{ar} < 1$. Prokaryotes are usually immortal, with $r_{ar} \leq 1$ whereas protozoa (unicellular eukaryotes) could be both immortal or mortal ($1 \leq r_{ar} \leq 1$), depending on environmental conditions and mode of reproduction (see Chap. 4). Prokaryotes and protozoa possess a variety of mechanisms that ensure immortality, thus clearly demonstrating that a cell has a huge potential of rejuvenation, not yet fully used in metazoans. In unfavorable conditions, age-related damage may be increased ($r_{ar} \geq 1$) and even a potentially immortal population can become mortal. Only in situations with $r_{ar} < 1$, it is reasonable to expect domination of rejuvenation.

11.3 Epigenetics and Biological Aging Clocks: Cytosine Methylation

Methylation of cytosine adjacent to guanine (CpG dinucleotide) is the main factor of epigenetic regulation. A cytosine methyl group projects into the major 'groove' of DNA double helix and affects binding of transcription factors. DNA methylation in aging is an example of how nature can solve complex problems by a relatively simple modification. A small fraction of the CpG sites is either hypermethylated or hypomethylated. Cytosines methylation occurs almost exclusively in the CpG islands. Aging occurs in a predictable manner—cytosine methylation is increased in gene promoters and is decreased in other parts of DNA. Promoter-resident CpGs which are commonly weakly methylated in young mice become hypermethylated with advanced age, while CpGs in gene-body or intergenic regions are initially higher methylated but turn hypomethylated in old animals (Jeong et al. 2023).

Table 11.1 Rejuvenation interventions evaluated by epigenetic clocks

Treatment	Species/sex/#	Age at the beginning of treatment	Duration of treatment	Test (tissue)	Effect	References
Vitamin D3	Humans 51 obese individuals: 43 ♀and 8 ♂	26.1 ± 9.3 years	16 weeks	Two epigenetic clocks (blood): DNAm age (Horvath 2013); DNAm age H (Hannum et al. 2013)	An average difference between epigenetic and chronological ages was ca. –1.7 years at the end of treatment	Chen et al. (2019)
rhGH + DHEA + Metformin	Humans 10 healthy ♂	51–65 years	12 months	Four epigenetic clocks (blood): DNAm age (Horvath 2013); DNAm age H (Hannum et al. 2013); DNAm PhenoAge (Levine et al. 2018); GrimAge (Lu et al. 2019)	An average difference between epigenetic and chronological ages was ca. –2.5 years at the end of treatment *Note:* six months after cessation of treatment, the difference between epigenetic and chronological ages was ca. –1.5 years	Fahy et al. (2019)
Intensive Relaxing Training	Humans 6 healthy individuals: 2 ♀and 4 ♂; 14 patients after MI: 3 ♀and 11 ♂	55.9 ± 6.1 years	2 months	DNAmAge was determined using Pyrosequencing® methodology (blood)	An average difference between epigenetic and chronological ages in healthy individuals was ca. –4.7 years at the end of treatment No effect on biological age of MI patients	Pavanello et al. (2019)

(continued)

Table 11.1 (continued)

Treatment	Species/sex/#	Age at the beginning of treatment	Duration of treatment	Test (tissue)	Effect	References
Mediterranean diet	Humans 2 populations (from Italy and Poland) of totally 120 healthy individuals: 69 ♀and 51 ♂	Italy: 33 ♀ 72.07 ± 3.80 years 27 ♂ 72.41 ± 3.91 years Poland: 36 ♀ 70.76 ± 4.01 years 24 ♂ 71.55 ± 4.27 years	One year	Two epigenetic clocks (blood): DNAm age (Horvath 2013); DNAm age H (Hannum et al. 2013)	An average difference between epigenetic and chronological ages in Polish individuals was ca. –1 year at the end of treatment; the effect was more prominent in females (ca. –2.5 years) No effect on biological age of Italian individuals	Gensous et al. (2020)
Rejuvant®	Humans 42 healthy individuals: 14 ♀and 28 ♂	♀ 64.1 years (43.5–72.5) Biological age: 62.2 ♂ 62.8 years (41.3–79.6) Biological age: 61.9	7 months in average (from 4 to 10 months)	TruMe Age test (saliva)	An average difference between epigenetic and chronological ages was ca. –7 years at the end of treatment *Note:* Rejuvant® is of additional benefit to chronologically and biologically older individuals	Demidenko et al. (2021)
Diet with supplemental probiotics and phytonutrients, exercises, sleep and relaxation	Humans 43 healthy ♂ (treatment group n = 18)	50–72 years	8 weeks	DNAmAge (saliva) (Horvath 2013)	Epigenetic age decreased by 2 years at the end of treatment	Fitzgerald et al. (2021)

(continued)

Table 11.1 (continued)

Treatment	Species/sex/#	Age at the beginning of treatment	Duration of treatment	Test (tissue)	Effect	References
Body Weight Reduction Program (BWRP)	Humans 72 obese individuals: 43 ♀and 29 ♂	51.5 ± 14.5 years	3 weeks	DNAm age (blood) (Zbieć-Piekarska et al. 2015)	BWRP reduced epigenetic age by ca. one year in the group with biological age higher than chronological age No effect on epigenetic age in the group with chronological age which did not exceed biological age	Rigamonti et al. (2022)
Progressive weighted wheel running (PoWeR)	C57BL/6 mice (♀n = 5)	22 months	8 weeks PoWeR after one introductory week	DNAge™ epigenetic aging clock	An average difference between epigenetic and chronological ages was ca. –4 months at the end of treatment	Murach et al. (2022)
Transient OSKM reprogramming	i4F-B mice with doxycycline-inducible OSKM transgen (n = 12)	6 ♀ 55 weeks 6 ♀+ ♂ 100 weeks	One week	*MspI*-based Reduced Representation Bisulfite Sequencing (pancreas, spleen and liver) (Meissner et al. 2005)	Partial epigenetic rejuvenation of multiple tissues Methylation profile and transcriptome of aged OSKM mice were placed between the young and aged groups	Chondronasiou et al. (2022)

(continued)

Table 11.1 (continued)

Treatment	Species/sex/#	Age at the beginning of treatment	Duration of treatment	Test (tissue)	Effect	References
Heterochronic parabiosis (HP)	C57BL/6 J mice (♀10 parabiotic pairs)	20 months	3 months of HP followed by a 2-month detachment period of anastomosed pairs	Five epigenetic clocks (blood, liver): Multi-tissue clocks (Meer et al. 2018; Thompson et al. 2018); Blood-specific clock (Petkovich et al. 2017); Maximum likelihood-based single-cell clock (Trapp et al. 2021); Mammalian methylation array (Lu et al. 2023)	An average difference between epigenetic and chronological ages was ca. –5 months immediately after 3 months of HP. The difference remains after 2-month detachment period of anastomosed pairs Old mice detached from heterochronic pairs lived longer than their isochronic controls: a 6-week extension in median life span and a 2-week extension in maximum life span	Zhang et al. (2023)

(continued)

Table 11.1 (continued)

Treatment	Species/sex/#	Age at the beginning of treatment	Duration of treatment	Test (tissue)	Effect	References
The exosome fraction (E5) of the porcine plasma from 6-month-old pigs	Sprague Dawley rats (♂ n = 18)	24 months	5 months	Six epigenetic clocks (blood, liver, heart and hypothalamus): Pan-tissue clock (rat); Brain clock (rat); Blood clock (rat); Liver (rat); Pan-tissue clock (human-rat, absolute age); Pan-tissue clock (human-rat, relative age)	An average difference between epigenetic and chronological ages was ca. –16 months at the end of treatment *Note* the treatment was accompanied by progressive improvement in the organ functions	Horvath et al. (2024)
Blood plasma from 2-month-old rats	Sprague Dawley rats (♀n = 47)	25.6 months	Once per week till natural death	Horvath Mammal Methyl Chip 320 assay (blood) (Arneson et al. 2022)	An average difference between epigenetic and chronological ages was ca. –12 months during treatment *Note* Increased mean lifespan by 2.2 months	Chiavellini et al. (2024)

rhGH, Recombinant human growth hormone; DHEA, Dehydroepiandrosterone; MI, myocardial infarction; OSKM, four Yamanaka's defined transcription factors

The efforts to estimate biological age have at least half-century history. Early versions of biological age tried to explain differences of aging trajectories by using simple multiple regression models including dozens of variables, with coefficients of R with chronological age usually lower than 0.5–0.7. Methylation clocks often show $R \geq 0.80–0.95$ across the entire age spectrum (from prenatals to centenarians). Epigenetic methylation clocks are more sensitive to longevity and mortality, i.e. to biological age, whereas transcriptomic, proteomic and metabolomic clocks show increased sensitivity to disease phenotypes (Rutledge et al. 2022; Zhang et al. 2022).

Almost twenty years ago, we suggested that common changes in the epigenetic control of gene expression may be one of the central mechanisms behind both aging and age-related diseases (ARDs) (Budovsky et al. 2006). Of note, the estimated loss of epigenetic control in aging may be one to two orders of magnitude greater than that of somatic DNA mutations (Bennett-Baker et al. 2003). Provided that epigenetics is a bridge between aging and ARDs, epigenetic reprogramming may serve a twofold purpose: (i) for extending the lifespan, and (ii) for curing ARDs. The idea of key position of epigenetics in aging/rejuvenation has conceptually been realized in epigenetic clocks giving researchers a powerful tool for estimating the biological age and aging rate, thus allowing to assess quantitatively the efficiency of rejuvenation factors. Nowadays, determining the biological age by epigenetic clocks has been becoming almost obligatory and routine procedure in rejuvenation studies (see Table 11.1), as well as for estimation of healthy status in diverse stressful and pathological conditions.

Epigenetic aging clock is focused on changes in DNA cytosine methylation in CpG dinucleotides. The first DNA methylation clock was built in 2011 by using about 100 saliva DNA samples (Bocklandt et al. 2011). In 2013, the 71-CpG clock was built by using 656 whole-blood samples (Hannum et al. 2013), followed by Horvath's pan-tissue 353-CpG clock using 8000 DNA samples and encompassing 51 different tissue and cell types (Horvath 2013). Since then, the concept of epigenetic clocks and its methodology have been rapidly developed, and most recently, using 11,754 methylation arrays, encompassing 59 tissue types across 185 mammalian species, a universal pan-mammalian biological clocks have been constructed (Lu et al. 2023). Most importantly, the clocks show unprecedented accuracy in correlation with chronological age ($R > 0.96$) and prediction of risk factors for health and mortality. Moreover, the clocks have become indispensable tools for assessment of rejuvenation and anti-aging (Ryan 2021; Tangili et al. 2023).

Transcriptomic/proteinomic clocks use data of gene expression analysis. The first transcriptomic clock was built in 2015 using the analysis of the peripheral blood mononuclear cell gene expression (Peters et al. 2015). Aging clocks based on single-cell transcriptomics can characterize cell-type-specific aging and rejuvenation (Buckley et al. 2023). In general, the transcriptomic and proteinomic clocks reveal high variability in age prediction and are less accurate than epigenetic clocks.

Biological clocks passed serious testing and trials by revealing an accelerated aging rate in the well-known pro-aging models of Down and Werner syndromes (Horvath et al. 2015; Maierhofer et al. 2017), Alzheimer and Parkinson diseases (Levine et al. 2015; Horvath and Ritz 2015), obesity and alcoholism (Horvath et al.

2014), cancer and osteoarthritis (Durso et al. 2017; Vidal-Bralo et al. 2016). Epigenetic clocks were also validated in health/longevity-promoting studies, thus showing that a decrease in biological age, i.e., rejuvenation, is achievable (see Table 11.1) (Chiavellini et al. 2021; Zhang et al. 2022; Johnson et al. 2022). Notably, a natural rejuvenation during early stages of embryogenesis was recently shown in mice (Kerepesi et al. 2021) and humans (Kerepesi and Gladyshev 2023).

Most biological clocks were built by linear regression methods which became the standard in the field (Rutledge et al. 2022; Zhang et al. 2022). They showed a high correlation with chronological age. However, the standard epigenetic clocks underestimate chronological ages of people 90+ (Dec et al. 2023; Guvatova et al. 2023). This apparent inconsistency could indicate that biological age of long-lived humans is actually much younger. Recently, using the machine learning-based methods, Dec et al. (2023) built the centenarian clock with high age correlations in the 80+ group, including centenarians and supercentenarians.

11.4 Epigenome and Decision-Making Structures

Genome has an efficient machinery for multiplying its information by replication and transcription. However, it is only a tool which starts working only after activation of epigenome. That said, a new metazoan organism emerges not with zygote but rather during development of epigenome in embryonic stem cells.

Epigenome DNA methylation and histone modifications (acetylation, phosphorylation, methylation, ubiquitination, sumoylation, etc.) are robust regulators of transcription and have memory for not only dynamic gene regulation but long-lasting decisions. Cells can remember their past experiences long after the signals fade away by using methyl tags (D'Urso and Brickner 2017). However, we posit that epigenome memory is hardly enough for comprehensive analysis and decision-making. The cooperation of epigenome with the hypothetical cellular analytical center (see Chap. 3) could be productive in resolving incessant chain of life collisions.

11.5 Concluding Remarks

Rejuvenology is a 'young' science with its own specific methodology and models, growing followers with sensible perspectives. Recent 'dragon jumping' in rejuvenation research has drawn attention of Academia and investors over reversal of aging rather than traditional decreasing the aging rate. The rejuvenation research gained unprecedented push due to construction of biological clocks which gave researchers a reliable tool for accurate estimation of biological age. In fact, assessment of age reversal by biological clocks is becoming a routine biomedical method. Biological age can be moved in both directions.

Ethics All the materials and data described in the book have been approved by Ethical Committee for Animal Experimentation or Helsinki Committee for Research on Humans of the organizations, where corresponding investigations have been done.

References

Arneson A, Haghani A, Thompson MJ, Pellegrini M, Kwon SB, Vu H, Maciejewski E, Yao M, Li CZ, Lu AT, Morselli M, Rubbi L, Barnes B, Hansen KD, Zhou W, Breeze CE, Ernst J, Horvath S (2022) A mammalian methylation array for profiling methylation levels at conserved sequences. Nat Commun 13(1):783. https://doi.org/10.1038/s41467-022-28355-z. PMID: 35145108

Bennett-Baker PE, Wilkowski J, Burke DT (2003) Age-associated activation of epigenetically repressed genes in the mouse. Genetics 165(4):2055–2062. https://doi.org/10.1093/genetics/165.4.2055. PMID: 14704185

Bocklandt S, Lin W, Sehl ME, Sánchez FJ, Sinsheimer JS, Horvath S, Vilain E (2011) Epigenetic predictor of age. PLoS ONE 6(6):e14821. https://doi.org/10.1371/journal.pone.0014821. PMID: 21731603

Buckley MT, Sun ED, George BM, Liu L, Schaum N, Xu L, Reyes JM, Goodell MA, Weissman IL, Wyss-Coray T, Rando TA, Brunet A (2023) Cell-type-specific aging clocks to quantify aging and rejuvenation in neurogenic regions of the brain. Nat Aging 3(1):121–137. https://doi.org/10.1038/s43587-022-00335-4. PMID: 37118510

Budovsky A, Muradian KK, Fraifeld VE (2006) From disease-oriented to aging/longevity-oriented studies. Rejuvenation Res 9(2):207–210. https://doi.org/10.1089/rej.2006.9.207. PMID: 16706644

Chen L, Dong Y, Bhagatwala J, Raed A, Huang Y, Zhu H (2019) Effects of Vitamin D3 Supplementation on Epigenetic Aging in Overweight and Obese African Americans With Suboptimal Vitamin D Status: A Randomized Clinical Trial. J Gerontol A Biol Sci Med Sci 74(1):91–98. https://doi.org/10.1093/gerona/gly223. PMID: 30256915

Chiavellini P, Canatelli-Mallat M, Lehmann M, Gallardo MD, Herenu CB, Cordeiro JL, Clement J, Goya RG (2021) Aging and rejuvenation—a modular epigenome model. Aging (Albany NY) 13(4):4734–4746. https://doi.org/10.18632/aging.202712. PMID: 33627519

Chiavellini P, Lehmann M, Gallardo MD, Canatelli Mallat M, Pasquini DC, Zoller JA, Gordevicius J, Girard M, Lacunza E, Herenu CB, Horvath S, Goya RG (2024) Young plasma rejuvenates blood dna methylation profile, extends mean lifespan and improves physical appearance in old rats. J Gerontol A Biol Sci Med Sci 2:glae071. https://doi.org/10.1093/gerona/glae071. PMID: 38430547

Chondronasiou D, Gill D, Mosteiro L, Urdinguio RG, Berenguer-Llergo A, Aguilera M, Durand S, Aprahamian F, Nirmalathasan N, Abad M, Martin-Herranz DE, Stephan-Otto Attolini C, Prats N, Kroemer G, Fraga MF, Reik W, Serrano M (2022) Multi-omic rejuvenation of naturally aged tissues by a single cycle of transient reprogramming. Aging Cell 21(3):e13578. https://doi.org/10.1111/acel.13578. PMID: 35235716

Dec E, Clement J, Cheng K, Church GM, Fossel MB, Rehkopf DH, Rosero-Bixby L, Kobor MS, Lin DT, Lu AT, Fei Z, Guo W, Chew YC, Yang X, Putra SED, Reiner AP, Correa A, Vilalta A, Pirazzini C, Passarino G, Monti D, Arosio B, Garagnani P, Franceschi C, Horvath S (2023) Centenarian clocks: epigenetic clocks for validating claims of exceptional longevity. Geroscience 45(3):1817–1835. https://doi.org/10.1007/s11357-023-00731-7. PMID: 36964402

Demidenko O, Barardo D, Budovskii V, Finnemore R, Palmer FR, Kennedy BK, Budovskaya YV (2021) Rejuvant®, a potential life-extending compound formulation with alpha-ketoglutarate and vitamins, conferred an average 8 year reduction in biological aging, after an average of 7 months of use, in the TruAge DNA methylation test. Aging (Albany NY) 13(22):24485–24499. https://doi.org/10.18632/aging.203736. PMID: 34847066

Damri M, Granot G, Ben-Meir H, Avivi Y, Plaschkes I, Chalifa-Caspi V, Wolfson M, Fraifeld V, Grafi G (2009) Senescing cells share common features with dedifferentiating cells. Rejuvenation Res 12(6):435–443. https://doi.org/10.1089/rej.2009.0887. PMID: 20041737

Durso DF, Bacalini MG, Sala C, Pirazzini C, Marasco E, Bonafé M, do Valle ÍF, Gentilini D, Castellani G, Faria AMC, Franceschi C, Garagnani P, Nardini C (2017) Acceleration of leukocytes' epigenetic age as an early tumor and sex-specific marker of breast and colorectal cancer. Oncotarget 8(14):23237–23245. https://doi.org/10.18632/oncotarget.15573. PMID: 28423572

D'Urso A, Brickner JH (2017) Epigenetic transcriptional memory. Curr Genet 63(3):435–439. https://doi.org/10.1007/s00294-016-0661-8. PMID: 27807647

Fahy GM, Brooke RT, Watson JP, Good Z, Vasanawala SS, Maecker H, Leipold MD, Lin DTS, Kobor MS, Horvath S (2019) Reversal of epigenetic aging and immunosenescent trends in humans. Aging Cell 18(6):e13028. https://doi.org/10.1111/acel.13028. PMID: 31496122

Fitzgerald KN, Hodges R, Hanes D, Stack E, Cheishvili D, Szyf M, Henkel J, Twedt MW, Giannopoulou D, Herdell J, Logan S, Bradley R (2021) Potential reversal of epigenetic age using a diet and lifestyle intervention: a pilot randomized clinical trial. Aging (Albany NY) 13(7):9419–9432. https://doi.org/10.18632/aging.202913. PMID: 33844651

Gensous N, Garagnani P, Santoro A, Giuliani C, Ostan R, Fabbri C, Milazzo M, Gentilini D, di Blasio AM, Pietruszka B, Madej D, Bialecka-Debek A, Brzozowska A, Franceschi C, Bacalini MG (2020) One-year Mediterranean diet promotes epigenetic rejuvenation with country- and sex-specific effects: a pilot study from the NU-AGE project. Geroscience 42(2):687–701. https://doi.org/10.1007/s11357-019-00149-0. PMID: 31981007

Gill D, Parry A, Santos F, Okkenhaug H, Todd CD, Hernando-Herraez I, Stubbs TM, Milagre I, Reik W (2022) Multi-omic rejuvenation of human cells by maturation phase transient reprogramming. Elife 11:e71624. https://doi.org/10.7554/eLife.71624. PMID: 35390271

Guvatova ZG, Kobelyatskaya AA, Pudova EA, Tarasova IV, Kudryavtseva AV, Tkacheva ON, Strazhesko ID, Moskalev AA (2023) Decelerated epigenetic aging in long livers. Int J Mol Sci 24(23):16867. https://doi.org/10.3390/ijms242316867. PMID: 38069189

Hannum G, Guinney J, Zhao L, Zhang L, Hughes G, Sadda S, Klotzle B, Bibikova M, Fan JB, Gao Y, Deconde R, Chen M, Rajapakse I, Friend S, Ideker T, Zhang K (2013) Genome-wide methylation profiles reveal quantitative views of human aging rates. Mol Cell 49(2):359–367. https://doi.org/10.1016/j.molcel.2012.10.016. PMID: 23177740

Horvath S (2013) DNA methylation age of human tissues and cell types. Genome Biol 14(10):R115. https://doi.org/10.1186/gb-2013-14-10-r115.Erratum.In:GenomeBiol.2015;16:96. PMID: 24138928

Horvath S, Ritz BR (2015) Increased epigenetic age and granulocyte counts in the blood of Parkinson's disease patients. Aging (Albany NY) 7(12):1130–1142. https://doi.org/10.18632/aging.100859. PMID: 26655927

Horvath S, Erhart W, Brosch M, Ammerpohl O, von Schönfels W, Ahrens M, Heits N, Bell JT, Tsai PC, Spector TD, Deloukas P, Siebert R, Sipos B, Becker T, Röcken C, Schafmayer C, Hampe J (2014) Obesity accelerates epigenetic aging of human liver. Proc Natl Acad Sci USA 111(43):15538–15543. https://doi.org/10.1073/pnas.1412759111. PMID: 25313081

Horvath S, Garagnani P, Bacalini MG, Pirazzini C, Salvioli S, Gentilini D, Di Blasio AM, Giuliani C, Tung S, Vinters HV, Franceschi C (2015) Accelerated epigenetic aging in down syndrome. Aging Cell 14(3):491–495. https://doi.org/10.1111/acel.12325. PMID: 25678027

Horvath S, Singh K, Raj K, Khairnar SI, Sanghavi A, Shrivastava A, Zoller JA, Li CZ, Herenu CB, Canatelli-Mallat M, Lehmann M, Habazin S, Novokmet M, Vučković F, Solberg Woods LC, Martinez AG, Wang T, Chiavellini P, Levine AJ, Chen H, Brooke RT, Gordevicius J, Lauc G, Goya RG, Katcher HL (2024) Reversal of biological age in multiple rat organs by young porcine plasma fraction. Geroscience 46(1):367–394. https://doi.org/10.1007/s11357-023-00980-6. PMID: 37875652

Jeong S, Cho S, Yang SK, Oh SA, Kang YK (2023) Parallel shift of DNA methylation and gene expression toward the mean in mouse spleen with aging. Aging (Albany NY) 15(14):6690–6709. https://doi.org/10.18632/aging.204903. PMID: 37494662

Ji S, Xiong M, Chen H, Liu Y, Zhou L, Hong Y, Wang M, Wang C, Fu X, Sun X (2023) Cellular rejuvenation: molecular mechanisms and potential therapeutic interventions for diseases. Signal Transduct Target Ther 8(1):116. https://doi.org/10.1038/s41392-023-01343-5. PMID: 36918530

Johnson AA, Torosin NS, Shokhirev MN, Cuellar TL (2022) A set of common buccal CpGs that predict epigenetic age and associate with lifespan-regulating genes. iScience 25(11):105304. https://doi.org/10.1016/j.isci.2022.105304. PMID: 36304118

Kerepesi C, Gladyshev VN (2023) Intersection clock reveals a rejuvenation event during human embryogenesis. Aging Cell 22(10):e13922. https://doi.org/10.1111/acel.13922. PMID: 37786333

Kerepesi C, Zhang B, Lee SG, Trapp A, Gladyshev VN (2021) Epigenetic clocks reveal a rejuvenation event during embryogenesis followed by aging. Sci Adv 7(26):eabg6082. https://doi.org/10.1126/sciadv.abg6082. PMID: 34172448

Levine ME, Lu AT, Bennett DA, Horvath S (2015) Epigenetic age of the pre-frontal cortex is associated with neuritic plaques, amyloid load, and Alzheimer's disease related cognitive functioning. Aging (Albany NY) 7(12):1198–1211. https://doi.org/10.18632/aging.100864. PMID: 26684672

Levine ME, Lu AT, Quach A, Chen BH, Assimes TL, Bandinelli S, Hou L, Baccarelli AA, Stewart JD, Li Y, Whitsel EA, Wilson JG, Reiner AP, Aviv A, Lohman K, Liu Y, Ferrucci L, Horvath S (2018) An epigenetic biomarker of aging for lifespan and healthspan. Aging (Albany NY) 10(4):573–591. https://doi.org/10.18632/aging.101414. PMID: 29676998

Lu AT, Quach A, Wilson JG, Reiner AP, Aviv A, Raj K, Hou L, Baccarelli AA, Li Y, Stewart JD, Whitsel EA, Assimes TL, Ferrucci L, Horvath S (2019) DNA methylation GrimAge strongly predicts lifespan and healthspan. Aging (Albany NY) 11(2):303–327. https://doi.org/10.18632/aging.101684. PMID: 30669119

Lu AT, Fei Z, Haghani A, Robeck TR, Zoller JA, Li CZ, Lowe R, Yan Q, Zhang J, Vu H, Ablaeva J, Acosta-Rodriguez VA, Adams DM, Almunia J, Aloysius A, Ardehali R, Arneson A, Baker CS, Banks G, Belov K, Bennett NC, Black P, Blumstein DT, Bors EK, Breeze CE, Brooke RT, Brown JL, Carter GG, Caulton A, Cavin JM, Chakrabarti L, Chatzistamou I, Chen H, Cheng K, Chiavellini P, Choi OW, Clarke SM, Cooper LN, Cossette ML, Day J, DeYoung J, DiRocco S, Dold C, Ehmke EE, Emmons CK, Emmrich S, Erbay E, Erlacher-Reid C, Faulkes CG, Ferguson SH, Finno CJ, Flower JE, Gaillard JM, Garde E, Gerber L, Gladyshev VN, Gorbunova V, Goya RG, Grant MJ, Green CB, Hales EN, Hanson MB, Hart DW, Haulena M, Herrick K, Hogan AN, Hogg CJ, Hore TA, Huang T, Izpisua Belmonte JC, Jasinska AJ, Jones G, Jourdain E, Kashpur O, Katcher H, Katsumata E, Kaza V, Kiaris H, Kobor MS, Kordowitzki P, Koski WR, Krützen M, Kwon SB, Larison B, Lee SG, Lehmann M, Lemaitre JF, Levine AJ, Li C, Li X, Lim AR, Lin DTS, Lindemann DM, Little TJ, Macoretta N, Maddox D, Matkin CO, Mattison JA, McClure M, Mergl J, Meudt JJ, Montano GA, Mozhui K, Munshi-South J, Naderi A, Nagy M, Narayan P, Nathanielsz PW, Nguyen NB, Niehrs C, O'Brien JK, O'Tierney Ginn P, Odom DT, Ophir AG, Osborn S, Ostrander EA, Parsons KM, Paul KC, Pellegrini M, Peters KJ, Pedersen AB, Petersen JL, Pietersen DW, Pinho GM, Plassais J, Poganik JR, Prado NA, Reddy P, Rey B, Ritz BR, Robbins J, Rodriguez M, Russell J, Rydkina E, Sailer LL, Salmon AB, Sanghavi A, Schachtschneider KM, Schmitt D, Schmitt T, Schomacher L, Schook LB, Sears KE, Seifert AW, Seluanov A, Shafer ABA, Shanmuganayagam D, Shindyapina AV, Simmons M, Singh K, Sinha I, Slone J, Snell RG, Soltanmaohammadi E, Spangler ML, Spriggs MC, Staggs L, Stedman N, Steinman KJ, Stewart DT, Sugrue VJ, Szladovits B, Takahashi JS, Takasugi M, Teeling EC, Thompson MJ, Van Bonn B, Vernes SC, Villar D, Vinters HV, Wallingford MC, Wang N, Wayne RK, Wilkinson GS, Williams CK, Williams RW, Yang XW, Yao M, Young BG, Zhang B, Zhang Z, Zhao P, Zhao Y, Zhou W, Zimmermann J, Ernst J, Raj K, Horvath S (2023) Universal DNA methylation age across mammalian tissues. Nat Aging 3(9):1144–1166. https://doi.org/10.1038/s43587-023-00462-6.Erratum.In:NatAging.2023. PMID: 37563227

Maierhofer A, Flunkert J, Oshima J, Martin GM, Haaf T, Horvath S (2017) Accelerated epigenetic aging in Werner syndrome. Aging (Albany NY) 9(4):1143–1152. https://doi.org/10.18632/aging.101217. PMID: 28377537

Meer MV, Podolskiy DI, Tyshkovskiy A, Gladyshev VN (2018) A whole lifespan mouse multi-tissue DNA methylation clock. Elife 7:e40675. https://doi.org/10.7554/eLife.40675. PMID: 30427307

Meissner A, Gnirke A, Bell GW, Ramsahoye B, Lander ES, Jaenisch R (2005) Reduced representation bisulfite sequencing for comparative high-resolution DNA methylation analysis. Nucleic Acids Res 33(18):5868–5877. https://doi.org/10.1093/nar/gki901. PMID: 16224102

Murach KA, Dimet-Wiley AL, Wen Y, Brightwell CR, Latham CM, Dungan CM, Fry CS, Watowich SJ (2022) Late-life exercise mitigates skeletal muscle epigenetic aging. Aging Cell 21(1):e13527. https://doi.org/10.1111/acel.13527. PMID: 34932867

Pavanello S, Campisi M, Tona F, Lin CD, Iliceto S (2019) Exploring epigenetic age in response to intensive relaxing training: a pilot study to slow down biological age. Int J Environ Res Public Health 16(17):3074. https://doi.org/10.3390/ijerph16173074. PMID: 31450859

Peters MJ, Joehanes R, Pilling LC, Schurmann C, Conneely KN, Powell J, Reinmaa E, Sutphin GL, Zhernakova A, Schramm K, Wilson YA, Kobes S, Tukiainen T; NABEC/UKBEC Consortium; Ramos YF, Göring HH, Fornage M, Liu Y, Gharib SA, Stranger BE, De Jager PL, Aviv A, Levy D, Murabito JM, Munson PJ, Huan T, Hofman A, Uitterlinden AG, Rivadeneira F, van Rooij J, Stolk L, Broer L, Verbiest MM, Jhamai M, Arp P, Metspalu A, Tserel L, Milani L, Samani NJ, Peterson P, Kasela S, Codd V, Peters A, Ward-Caviness CK, Herder C, Waldenberger M, Roden M, Singmann P, Zeilinger S, Illig T, Homuth G, Grabe HJ, Völzke H, Steil L, Kocher T, Murray A, Melzer D, Yaghootkar H, Bandinelli S, Moses EK, Kent JW, Curran JE, Johnson MP, Williams-Blangero S, Westra HJ, McRae AF, Smith JA, Kardia SL, Hovatta I, Perola M, Ripatti S, Salomaa V, Henders AK, Martin NG, Smith AK, Mehta D, Binder EB, Nylocks KM, Kennedy EM, Klengel T, Ding J, Suchy-Dicey AM, Enquobahrie DA, Brody J, Rotter JI, Chen YD, Houwing-Duistermaat J, Kloppenburg M, Slagboom PE, Helmer Q, den Hollander W, Bean S, Raj T, Bakhshi N, Wang QP, Oyston LJ, Psaty BM, Tracy RP, Montgomery GW, Turner ST, Blangero J, Meulenbelt I, Ressler KJ, Yang J, Franke L, Kettunen J, Visscher PM, Neely GG, Korstanje R, Hanson RL, Prokisch H, Ferrucci L, Esko T, Teumer A, van Meurs JB, Johnson AD (2015) The transcriptional landscape of age in human peripheral blood. Nat Commun. https://doi.org/10.1038/ncomms9570. PMID: 26490707

Petkovich DA, Podolskiy DI, Lobanov AV, Lee SG, Miller RA, Gladyshev VN (2017) Using DNA Methylation Profiling to Evaluate Biological Age and Longevity Interventions. Cell Metab 25(4):954–960.e6. https://doi.org/10.1016/j.cmet.2017.03.016. PMID: 28380383

Poganik JR, Zhang B, Baht GS, Tyshkovskiy A, Deik A, Kerepesi C, Yim SH, Lu AT, Haghani A, Gong T, Hedman AM, Andolf E, Pershagen G, Almqvist C, Clish CB, Horvath S, White JP, Gladyshev VN (2023) Biological age is increased by stress and restored upon recovery. Cell Metab 35(5):807-820.e5. https://doi.org/10.1016/j.cmet.2023.03.015. PMID: 37086720

Puri D, Wagner W (2023) Epigenetic rejuvenation by partial reprogramming. Bioessays 45(4):e2200208. https://doi.org/10.1002/bies.202200208. PMID: 36871150

Rigamonti AE, Bollati V, Favero C, Albetti B, Caroli D, Abbruzzese L, Cella SG, Sartorio A (2022) Effect of a 3-week multidisciplinary body weight reduction program on the epigenetic age acceleration in obese adults. J Clin Med 11(16):4677. https://doi.org/10.3390/jcm11164677. PMID: 36012914

Rutledge J, Oh H, Wyss-Coray T (2022) Measuring biological age using omics data. Nat Rev Genet 23(12):715–727. https://doi.org/10.1038/s41576-022-00511-7. PMID: 35715611

Ryan CP (2021) Epigenetic clocks": Theory and applications in human biology. Am J Hum Biol 33(3):e23488. https://doi.org/10.1002/ajhb.23488. PMID: 32845048

Simpson DJ, Olova NN, Chandra T (2021) Cellular reprogramming and epigenetic rejuvenation. Clin Epigenetics 13(1):170. https://doi.org/10.1186/s13148-021-01158-7. PMID: 34488874

Tangili M, Slettenhaar AJ, Sudyka J, Dugdale HL, Pen I, Palsbøll PJ, Verhulst S (2023) DNA methylation markers of age(ing) in non-model animals. Mol Ecol 32(17):4725–4741. https://doi.org/10.1111/mec.17065. PMID: 37401200

Thompson MJ, Chwiałkowska K, Rubbi L, Lusis AJ, Davis RC, Srivastava A, Korstanje R, Churchill GA, Horvath S, Pellegrini M (2018) A multi-tissue full lifespan epigenetic clock for mice. Aging (Albany NY) 10(10):2832–2854. https://doi.org/10.18632/aging.101590. PMID: 30348905

Trapp A, Kerepesi C, Gladyshev VN (2021) Profiling epigenetic age in single cells. Nat Aging 1(12):1189–1201. https://doi.org/10.1038/s43587-021-00134-3. PMID: 36211119

Vidal-Bralo L, Lopez-Golan Y, Mera-Varela A, Rego-Perez I, Horvath S, Zhang Y, Del Real Á, Zhai G, Blanco FJ, Riancho JA, Gomez-Reino JJ, Gonzalez A (2016) Specific premature epigenetic aging of cartilage in osteoarthritis. Aging (Albany NY) 8(9):2222–2231. https://doi.org/10.18632/aging.101053. PMID: 27689435

Yang JH, Petty CA, Dixon-McDougall T, Lopez MV, Tyshkovskiy A, Maybury-Lewis S, Tian X, Ibrahim N, Chen Z, Griffin PT, Arnold M, Li J, Martinez OA, Behn A, Rogers-Hammond R, Angeli S, Gladyshev VN, Sinclair DA (2023) Chemically induced reprogramming to reverse cellular aging. Aging (Albany NY) 15(13):5966–5989. https://doi.org/10.18632/aging.204896. PMID: 37437248

Zbieć-Piekarska R, Spólnicka M, Kupiec T, Parys-Proszek A, Makowska Ż, Pałeczka A, Kucharczyk K, Płoski R, Branicki W (2015) Development of a forensically useful age prediction method based on DNA methylation analysis. Forensic Sci Int Genet 17:173–179. https://doi.org/10.1016/j.fsigen.2015.05.001. PMID: 26026729

Zhang B, Trapp A, Kerepesi C, Gladyshev VN (2022) Emerging rejuvenation strategies–reducing the biological age. Aging Cell 21(1):e13538. https://doi.org/10.1111/acel.13538. PMID: 34972247

Zhang B, Lee DE, Trapp A, Tyshkovskiy A, Lu AT, Bareja A, Kerepesi C, McKay LK, Shindyapina AV, Dmitriev SE, Baht GS, Horvath S, Gladyshev VN, White JP (2023) Multi-omic rejuvenation and life span extension on exposure to youthful circulation. Nat Aging. 3(8):948–964. https://doi.org/10.1038/s43587-023-00451-9. PMID: 37500973

Chapter 12
Rejuvenation Is on Its Way

Abstract Heterochronic parabiosis or blood plasma exchange has lately become popular models in rejuvenation research because they demonstrate systemic decrease of biological age in a variety of tissues. Partial dilution of blood plasma can be equally efficient. We suggest the rejuvenation model of inter/intra cellular washup for removal of damaged cellular constitutes. There are various forms of adult stem cells, including pluripotent stem cells, and proliferatively active zones in the brain, muscles, heart, liver, kidney, and other tissues. These systems will obviously become the active sites in rejuvenation. The important point is that a cell 'knows' how to find the way back to pluripotency and rejuvenation. Rejuvenation is a natural process. The major natural rejuvenative events in mammals occur in meiosis/gametogenesis and early embryonic development, aimed at ridding off the load of aging. Rejuvenation can occur in adult life as well, for example, via cellular reprogramming.

Keywords Blood exchange models · Cellular washup · Rejuvenation of organs · First and second laws of rejuvenation · Rejuvenation milestones before birth

12.1 Rejuvenation and Blood Exchange Models in Mammals

Heterochronic parabiosis (HP) and blood/plasma exchange became popular models in rejuvenation research. In fact, blood has been used for rejuvenation since ancient times. Blood exchange and transfusion are usual biomedical procedures. The classical HP model, using surgically connected young and old rats with sharing blood circulation, was first suggested by McCay in 1957 (McCay et al. 1957). Since then, HP and blood exchange have been successfully tested in dozens of experiments (Conboy et al. 2005, 2013; Ma et al. 2022; Pálovics et al. 2022; Buckley et al. 2023; Zhang et al. 2023; Horvath et al. 2024) (see also Chap. 11, Table 11.1), with only few exceptions (Shytikov et al. 2014; Yankova et al. 2022). After a 3-months HP, along with improved physiological parameters, transcriptomic and epigenomic profiling of the old mouse partner revealed a global multi-omics rejuvenation effect.

K. K. Muradian and V. E. Fraifeld, *Rejuvenation and Longevity*, Healthy Ageing and Longevity 20, https://doi.org/10.1007/978-3-031-64995-0_12

Heterochronic parabiosis extended health and life span of the old mice, presumably due to a long-term epigenetic and transcriptome reorganization (Zhang et al. 2023). Heterochronic parabiosis and systemic administration of young blood plasma into aged mice showed that exposure of aged animals to young blood can reverse the pre-existing deficits in aging hippocampus and brain (Villeda et al. 2014). Comparison of young and old animal plasma identified platelet factor 4 (PF4), which could be responsible for attenuating age-related neuroinflammation (Schroer et al. 2023). Systemic administration of PF4 improved neurogenesis in the hippocampus and cognitive function in aged mice (Leiter et al. 2023). Although PF4 is involved in myriad processes, it could be only a small piece of the larger puzzle known as brain rejuvenation (Izquierdo, 2024).

The transfusion of whole blood, blood plasma or its fractions as well as cerebrospinal fluid (CSF) from young to old individuals were tested in several rodent models and in humans with Alzheimer's disease (Table 12.1).

As seen in Table 12.1, the results of 'young' blood transfusion or its components were consistently beneficial for old individuals. Interestingly, this effect was preserved even when transfusion was conducted between different mammalian species. For example, transfusion of human umbilical cord blood plasma to middle-aged mice significantly improved murine bran functions (Castellano et al. 2017). In experiments carried out by Horvath and colleagues (2024), the exosome fraction of blood plasma from young adult pigs also rejuvenated old rat tissues at the epigenetic level in an 'interspecies' manner. The treatment demonstrated that young porcine plasma reversed aging in rats, confirmed by epigenetic clocks, IgG glycans, and other biomarkers of aging (Horvath et al. 2024).

After successful study of HP models, Conboy and colleagues addressed a principal question: are young blood factors necessary for a multi-tissue rejuvenation? Using only a single procedure of half a blood plasma exchange by saline-albumin appeared suffice to exceed the rejuvenative effects of HP on liver adiposity or fibrosis and increasing hippocampal neurogenesis in old mice (Mehdipour et al. 2020, 2021). Moreover, the self-sufficiency of plasma dilution for rejuvenation was shown in people as well (Kiprov et al. 2021; Kim et al. 2022). These data suggest that rejuvenative effects could be achieved through dilution of age-related systemic factors.

12.2 Cellular Washup

The bacterial immortality is often explained by removal of damaged cellular constituents through replicative dilution and excretion of exocellular vesicles (Sheldrake, 2022). For multicellular animals, this way encounters principal obstacles, because most of cells in metazoans are post-replicative and the excretion of exocellular vesicles often provokes unhealthy inflammaging environment.

Cellular hydration could be stimulated by increasing water content in the blood and other body's circulating fluids. To some extent, it could be achieved by drinking water.

Table 12.1 Blood/plasma transfusion from young to old individuals

Species (sex/#)	Treatment/age of donors	Age at the beginning of treatment	Duration of treatment	Treatment procedure	Effect	References
Alzheimer's disease patients (♂ n = 6; ♀n = 12)	Fresh frozen plasma from young healthy human male donors 18 to 30 years of age	50 to 90 years	4 weeks	250 ml of young fresh frozen plasma i.v	Improvement on the Functional Activities Questionnaire and the Alzheimer's Disease Cooperative Study Activities of Daily Living Inventory in mild cognitive impairment after plasma treatment	Sha et al. (2019)
C57BL/6 J mice (♂ n = 8)	Blood plasma from 3-month-old mice	18 months	24 days	100 μl of "young" plasma i.v. every third day (8 times)	Improved hippocampal-dependent learning and memory; Rejuvenated synaptic plasticity; Activation of the cyclic AMP response element binding protein (Creb)	Villeda et al. (2014)

(continued)

Table 12.1 (continued)

Species (sex/#)	Treatment/age of donors	Age at the beginning of treatment	Duration of treatment	Treatment procedure	Effect	References
C57BL/6 J mice (♂ n = 4)	Blood from 2-month-old mice	22 months	7 days (after a one-day procedur)	Heterochronic blood exchange (HP-like) 150 µl of whole blood was transferred from one mouse to another, 15 times with a 30 s delay between blood administration and withdrawal. The procedure was done twice within 24 h to yield a < 90% homogenization of blood between the two animals	Restoration of old hepatocyte proliferative rate to the 'young' level Rapid decline of fibrotic spots and adiposity in old livers	Rebo et al. (2016)
NSG mice (♂ n = 7)	Human umbilical cord blood plasma	13.5 months	2 weeks	175 µl of cord plasma i.v. 6–8 times every other day	Increased long-term potentiation; Improved memory	Castellano et al. (2017)
Sprague Dawley rats (♂ n = 4)	Blood plasma from 3-month-old rats	22 months	4 weeks	1 ml of "young" plasma i.v. 3 times per week	Improved hepatic functions; Improved hepatic regeneration after partial hepatectomy	Liu et al. (2018)
Wistar rats (n = 6)	Blood plasma from 2–3-month-old rats	24 months	4 weeks	1 ml of "young" plasma i.v. 3 times per week	Improved biomarkers of redox balance in blood	Tripathi et al. (2021)

(continued)

Table 12.1 (continued)

Species (sex/#)	Treatment/age of donors	Age at the beginning of treatment	Duration of treatment	Treatment procedure	Effect	References
C57BL/6 J mice (♂ n = 9)	Cerebrospinal fluid (CSF) from 2.5-month-old mice	18–22 months	7 days	Continuous i.c.v. infusion of 90 μl through the coiled catheter to the right lateral ventricle at a rate of 0.5 μl per hour	Improved memory; Increased proliferation of oligodendrocytes, astrocytes and microglia in hippocampus; Increased number of myelinated axons in hippocampus	Iram et al. (2022)
Sprague Dawley rats (♂ n = 18)	The exosome fraction (E5) of the porcine plasma from 6-month-old pigs	24 months	5 months	1.43 g of solid precipitate of exosomes per 500 g body weight Two series of 4 i.v. injections every alternate day for 8 days: First started at Day 1, Second started at Day 95	Progressive improvement in organ functions: a decrease in blood bilirubin, glucose, creatinine, cholesterol, pro-inflammatory cytokines to the levels of young animals; improvement of cognitive functions Reversal of biological age	Horvath et al. (2024)
Sprague Dawley rats (♀n = 47)	Blood plasma from 2-month-old rats	25.6 months	Till natural death	Slow (2 min) i.p. injection of 1 ml of "young" plasma once per week	Increase in mean lifespan by 2.2 months Biological age reversal	Chiavellini et al. (2024)
McGill-R-Thy1-APP transgenic rat model for Alzheimer's disease (♂ n = 23)	Blood plasma from 2.5–3.5-month-old Wistar rats after 6 weeks of exercise training	5 months *Note:* extracellular plaques emerge at 6 months of age	6 weeks	2 ml of "young" plasma per kg body weight i.v. 2 or 3 times per week; 14 injections in total	Promoted adult hippocampal neurogenesis	Norevik et al. (2024)

i.v.—intravenously; i.p.—intraperitoneally; i.c.v.—intracerebroventricular

We suggest, however, that dilution and clearance of unwanted inter- and intra-cellular matter could be enhanced by using the 'controlled washup'—the solutions containing the natural surface-active substances and organic liquids. Their application in optimal concentrations could enhance solubility of denatured protein-lipid complexes of the damaged structures.

12.3 Rejuvenation of Organs

Rapid rejuvenation of the whole organism seems still problematic. In the near future, successes in rejuvenation could rather be expected by the step-by-step and 'organ-by-organ' manner. Here, we consider in brief several organs which age-related changes were partially reversed by application of diverse interventions. Of note, at least three of them—blood exchange, exercise and diet—have been reported to impact aged stem cell function (Brunet et al. 2023).

Skin is the outer protective barrier and the largest human organ, often injured by both external and internal damaging factors. On the other hand, frequent wound healing makes skin an efficient organ for regeneration and rejuvenation. Remarkably, long-lived mouse strains preserve the rate of skin wound healing, thus pointing toward putative links between longevity and regenerative capacity (Yanai et al. 2011, 2015, 2016). Though differing in specific mechanisms, the basic events during skin repair have much in common across a variety of wounded organs (Gurtner et al. 2008). Epigenetics, DNA repair, apoptosis, telomere attrition, genome stability, mitochondrial ROS, transposons, vitamins, collagen and elastin, cell cycle control, and many other factors are in the focus of dermatological aging and rejuvenation (Gaur et al. 2017; Shin et al. 2023).

Adult stem cells of the bone marrow, adipose tissue and blood are the most available sources of the multipotent mesenchymal stem cells (MSCs) that have been recognized as therapeutic agents for skin regeneration and rejuvenation. MSCs are potent stimulators of collagen and elastic fibers production as well as skin vascularization (Charles-de-Sá et al. 2015; Jo et al. 2021). Also, the extracellular vesicles derived from adipose stem cells can decrease skin wrinkles, promote epidermal cell proliferation, decline macrophage infiltration and collagen degradation (Xu et al. 2020). Sagging and wrinkling make skin a visual demonstration of human age and well-being, written on the face. Skin rejuvenation is becoming a growing enterprise which brings pleasure to patients and money to the industry. The current worldwide annual expenditure for skin rejuvenation is around \$ US25 billion and will reach \$ US45 billion by 2030 (Griffiths et al. 2023).

Brain aging affects other organs in the body and contributes to age-related physiological decline. Neural stem cells of the subventricular zone of the lateral ventricles, the subgranular zone in the dentate gyrus of the hippocampus, and the neurogenic area of the hypothalamus generate new neurons throughout life. The reduced neurogenesis in old animals can be substantially restored by pharmacological agents (rapamycin), lifestyle factors (diet, exercise), or systemic blood changes (heterochronic parabiosis, administration of young blood plasma) (Katsimpardi et al. 2014; Bouchard and Villeda 2015; Navarro Negredo et al. 2020; Brunet et al. 2023; Buckley et al. 2023). Importantly, the positive changes in neurogenesis and regenerative capacity of neural stem cells were accompanied by the improvement of brain functions. Also, impaired neurogenesis and behavioral deficits could be efficiently treated with systemic or intracerebral transplantation of MSCs. In these cases, the engrafted MSCs did not differentiate into new neurons but rather support *endogenous neurogenesis* in a paracrine manner (Gobshtis et al. 2017, 2021; Tfilin et al. 2023). The neurogenic areas of the brain, as well as similar areas in other organs, can be important sources of cellular recovery in various rejuvenative procedures.

Heart failure in the elderly often involves cardiac hypertrophy, which could be attributed to the age-related loss of some blood-circulating factors. In support of this notion is an observation indicating that heterochronic parabiosis exhibited dramatic regression of cardiac hypertrophy in old mice, and this occurred due to the restored levels of the TGF-β superfamily member GDF11. Accordingly, treatment of intact old mice with GDF11 reversed cardiac hypertrophy (Loffredo et al. 2013). It seems reasonable to speculate that such rejuvenating factors exist for diverse tissues, and their identification—hopefully in the near future—would be a real breakthrough in the field (Mendelsohn and Larrick 2013; Nguyen and Sussman 2015).

Liver aging is often associated with elevated levels of hepatic enzymes, accumulation of lipids and lipofuscin, fibroproliferative changes, reduced regenerative capacity, and attenuated autophagy. Intravenous infusions of whole blood, plasma, or its exosome fraction from young to old rodents reversed the above age-related changes (Rebo et al. 2016; Liu et al. 2018; Horvath et al. 2024). 'Young' serum could protect the cultured old hepatocytes from cellular senescence (Xu et al. 2021; Liu et al. 2018).

Skeletal muscle aging is often a consequence of loss of adult residual stem cells (satellite cells). These cells are long-lived and can support muscle regeneration throughout life. Numerous in vivo and ex vivo interventions can restore regenerative potential of satellite cells. Positive effects of heterochronic parabiosis, caloric restriction, rapamycin, oxytocin, nicotinamide, senolytics, and in vivo reprogramming anticipate emerging field to store youthfulness of satellite cells and hopefully the muscles (Conboy et al. 2005; Sousa-Victor et al. 2020; Muñoz-Cánoves et al. 2020; Brunet et al. 2023). Even late-life exercise training can significantly attenuate epigenetic skeletal muscle aging (Murach et al. 2022).

12.4 The First and Second Laws of Rejuvenation

As a good soldier dreams to become a general, a differentiated cell obviously strives to become an immortal stem cell. In multicellular organisms, differentiated cells often do it 'illegally' by transforming into immortal cancer cells.

From the very protocells, a cell clearly demonstrates that it 'wants and knows' to be young and immortal, even in active resistance from the surrounding cells and numerous controlling factors. We suggest to name this irresistible ***aspiration of a cell to be young and immortal*** as the first law of rejuvenation. This basic capacity may be the main driving force of the infinity of biological life as such.

Protocells and prokaryotes originated much earlier than the differentiated cells. They had plenty possibilities to occupy thermodynamically favorable regulatory and metabolic landscapes. Due to such advantages, ***a cell can spontaneously roll down and occupy a defined most stable thermodynamic position***. We suggest to name this phenomenon—the second law of rejuvenation, in analogy with the second law of thermodynamics. The second law of thermodynamics is a universal law and is valid in all corners of the universe. Therefore, the same should be with the second law of rejuvenation: a cell can be young and immortal anytime and everywhere (see also Chap. 3).

12.5 Rejuvenation Before and After the Birth

Milestones of Rejuvenation

Rejuvenation is a natural process for ridding off the load of aging. The major rejuvenative events in multicellular organisms, mammals included, occur primarily in meiosis/gametogenesis and early embryonic development (Muradian et al. 2019; Gladyshev, 2021; Kerepesi et al. 2021; Kerepesi and Gladyshev 2023) (Fig. 12.1). After fertilization, the next rejuvenation milestone falls on the zygotic genome activation (ZGA). The third and the final major rejuvenative event starts from preimplanted blastocyst and ends at early implantation. Profound resetting of the epigenome with involvement of transposable elements are the hallmarks for all these stages (see Chap. 7).

In Chap. 8, we discussed in detail why hypoxic/hypercapnic environment is so important for ESCs and early embryogenesis in general. This is especially true for ESCs of the preimplanted blastocyst, when hypoxia/hypercapnia reaches its peak, thus minimizing oxidative damage and preventing the ESCs from premature differentiation. A few years ago, we wrote that "this short period of ESC life provides a unique opportunity for the 'last-minute' correction of the DNA damage, before ESCs start to differentiate. The very fact that the life still exists, and the accumulated damage does not pass on from generation to generation is, to a great extent, guaranteed by that 'last-minute' correction of ESCs" (Muradian et al. 2019). Our hypothesis gained a strong support from the studies of Prof. Gladyshev's group who showed that after the 'last-minute' correction, embryo reaches the minimal biological age ('ground zero') (Gladyshev 2021; Kerepesi et al. 2021; Kerepesi and Gladyshev 2023). Of note, most spontaneous embryo abortions (miscarriages) fall on the period between late preimplantation and early implantation (Muter et al. 2023), thus ensuring additional line of protection, if something went wrong.

Cell reprogramming and rejuvenation events in adult life

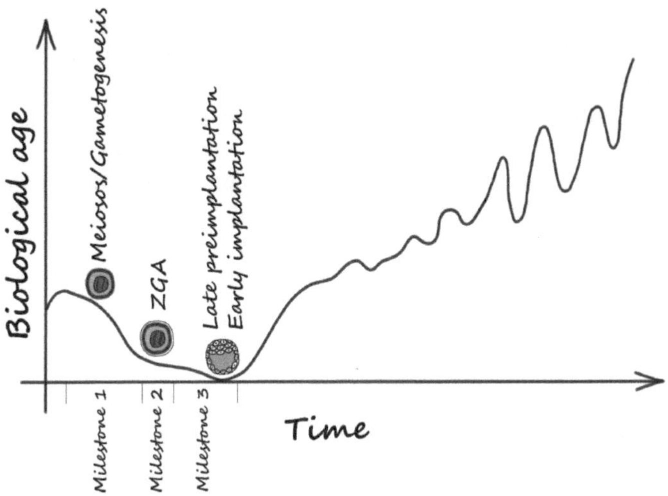

Fig. 12.1 Natural milestones of rejuvenation in mammals

If the somatic stem cells would be able to maintain their regenerative capacity over time, this might resolve rejuvenation issues. Unfortunately, the pool of adult stem cells is limited, and they undergo cell aging with a consequent loss of functionality. This in turn greatly affects the tissue regenerative capacity and eventually contributes to organismal aging. Cell reprogramming may, to some extent, serve as a solution. Moreover, it is not necessary to reach a pluripotent state, because partial reprogramming by limited exposure to reprogramming factors can reset epigenetic aging clocks without loss of cellular identity (Puri and Wagner 2023; Yücel and Gladyshev 2024).

The defined small molecules (SMs) and their cocktails able to induce or enhance cell reprogramming have been discovered and successfully tested in vitro, including human somatic cells (Guan et al. 2022; Yang et al. 2023). They can interact with the defined transcription factors by enhancing or substituting their activity. We have recently hypothesized that cell reprogramming could be a *natural* process triggered by two major networks: a network of transcription factors (TFs, e.g. Yamanaka's factors, OSKM) and a metabolite network (e.g. similar to defined SMs) (Muradian and Fraifeld 2020; Knyazer et al. 2021). Indeed, many SMs are found in the human metabolome (derivatives of nucleotides, fatty acids, etc.), and many more metabolites are functionally similar to SMs, thus offering the potential for cell reprogramming. Moreover, between TFs and SM cocktails, there are many common pathways but not specific targets (Knyazer et al. 2021), indicating that they may cooperate with each other, and thereby substantially increase the efficiency of cell reprogramming. In some conditions, the concentrations of SM-like natural metabolites could spontaneously reach levels sufficient for triggering cell reprogramming and partial rejuvenation. It is tempting to speculate that the probability of such combinations of natural SMs would be particularly higher in species with extreme longevity, thus explaining in part the very origin of this phenomenon. All in all, further investigation of SMs and their relationships with longevity regulators will be helpful for developing the most optimal SM cocktails for cell reprogramming with a perspective for rejuvenation and life span extension.

The very possibility to decrease the epigenetic clock in advanced age was shown by means of various interventions (see Chap. 11, Table 11.1). The rejuvenation effects could be even higher in chronologically and biologically older individuals or in individuals whose biological age is higher than chronological one (Demidenko et al. 2021; Rigamonti et al. 2022).

12.6 Concluding Remarks

Rejuvenology is a 'young' science with its own specific methodology and models, growing followers with sensible perspectives. Recent 'dragon jumping' in rejuvenation research has drawn attention of Academia and investors over reversal of aging rather than traditional decreasing the aging rate.

Based on the dilution principle widely used by immortal protozoa and metazoans with life-long body growth (see Chaps. 4 and 5), there are reasons to believe that inter- and intra-cellular dilution could be efficient in mammals and humans as well. Application of natural surface-active substances or organic liquids could be useful for dissolution of denaturized protein-lipid complexes and lipofuscin.

Skin could be an example of rejuvenation for other tissues and organs. Restoring the pool of adult stem cells, including adult pluripotent stem cells, seems a promising and achievable perspective for tissue/organ rejuvenation. Almost all tissues have specific active sources for cellular recovery, as are the neurogenic zones in the brain. These zones most possibly will become active sites of rejuvenation. Rejuvenation is on its way!

Ethics All the materials and data described in the book have been approved by Ethical Committee for Animal Experimentation or Helsinki Committee for Research on Humans of the organizations, where corresponding investigations have been done.

References

Bouchard J, Villeda SA (2015) Aging and brain rejuvenation as systemic events. J Neurochem 132(1):5–19. https://doi.org/10.1111/jnc.12969. PMID: 25327899

Brunet A, Goodell MA, Rando TA (2023) Ageing and rejuvenation of tissue stem cells and their niches. Nat Rev Mol Cell Biol 24(1):45–62. https://doi.org/10.1038/s41580-022-00510-w. PMID: 35859206

Buckley MT, Sun ED, George BM, Liu L, Schaum N, Xu L, Reyes JM, Goodell MA, Weissman IL, Wyss-Coray T, Rando TA, Brunet A (2023) Cell-type-specific aging clocks to quantify aging and rejuvenation in neurogenic regions of the brain. Nat Aging 3(1):121–137. https://doi.org/10.1038/s43587-022-00335-4. PMID: 37118510

Castellano JM, Mosher KI, Abbey RJ, McBride AA, James ML, Berdnik D, Shen JC, Zou B, Xie XS, Tingle M, Hinkson IV, Angst MS, Wyss-Coray T (2017) Human umbilical cord plasma proteins revitalize hippocampal function in aged mice. Nature 544(7651):488–492. https://doi.org/10.1038/nature22067. PMID: 28424512

Charles-de-Sá L, Gontijo-de-Amorim NF, Maeda Takiya C, Borojevic R, Benati D, Bernardi P, Sbarbati A, Rigotti G (2015) Antiaging treatment of the facial skin by fat graft and adipose-derived stem cells. Plast Reconstr Surg 35(4):999–1009. https://doi.org/10.1097/PRS.0000000000001123. PMID: 25811565

Chiavellini P, Lehmann M, Gallardo MD, Canatelli Mallat M, Pasquini DC, Zoller JA, Gordevicius J, Girard M, Lacunza E, Herenu CB, Horvath S, Goya RG (2024) Young plasma rejuvenates blood dna methylation profile, extends mean lifespan and improves physical appearance in old rats. J Gerontol A Biol Sci Med Sci 2:glae071. https://doi.org/10.1093/gerona/glae071. PMID: 38430547

Conboy IM, Conboy MJ, Wagers AJ, Girma ER, Weissman IL, Rando TA (2005) Rejuvenation of aged progenitor cells by exposure to a young systemic environment. Nature 433(7027):760–764. https://doi.org/10.1038/nature03260. PMID: 15716955

Conboy MJ, Conboy IM, Rando TA (2013) Heterochronic parabiosis: historical perspective and methodological considerations for studies of aging and longevity. Aging Cell 12(3):525–530. https://doi.org/10.1111/acel.12065. PMID: 23489470

Demidenko O, Barardo D, Budovskii V, Finnemore R, Palmer FR, Kennedy BK, Budovskaya YV (2021) Rejuvant®, a potential life-extending compound formulation with alpha-ketoglutarate and vitamins, conferred an average 8 year reduction in biological aging, after an average of 7 months of use, in the TruAge DNA methylation test. Aging (Albany NY). 13(22):24485–24499. https://doi.org/10.18632/aging.203736. PMID: 34847066

Gaur M, Dobke M, Lunyak VV (2017) Mesenchymal stem cells from adipose tissue in clinical applications for dermatological indications and skin aging. Int J Mol Sci 18(1):208. https://doi.org/10.3390/ijms18010208. PMID: 28117680

Gladyshev VN (2021) The ground zero of organismal life and aging. Trends Mol Med 27(1):11–19. https://doi.org/10.1016/j.molmed.2020.08.012. PMID: 32980264

Gobshtis N, Tfilin M, Wolfson M, Fraifeld VE, Turgeman G (2017) Transplantation of mesenchymal stem cells reverses behavioural deficits and impaired neurogenesis caused by prenatal exposure to valproic acid. Oncotarget 8(11):17443–17452. https://doi.org/10.18632/oncotarget.15245. PMID: 28407680

Gobshtis N, Tfilin M, Fraifeld VE, Turgeman G (2021) Transplantation of mesenchymal stem cells causes long-term alleviation of schizophrenia-like behaviour coupled with increased neurogenesis. Mol Psychiatry 26(8):4448–4463. https://doi.org/10.1038/s41380-019-0623-x. PMID: 31827249

Griffiths TW, Watson REB, Langton AK (2023) Skin ageing and topical rejuvenation strategies. Br J Dermatol 189(Suppl 1):i17–i23. https://doi.org/10.1093/bjd/ljad282. PMID: 37903073

Guan J, Wang G, Wang J, Zhang Z, Fu Y, Cheng L, Meng G, Lyu Y, Zhu J, Li Y, Wang Y, Liuyang S, Liu B, Yang Z, He H, Zhong X, Chen Q, Zhang X, Sun S, Lai W, Shi Y, Liu L, Wang L, Li C, Lu S, Deng H (2022) Chemical reprogramming of human somatic cells to pluripotent stem cells. Nature 605(7909):325–331. https://doi.org/10.1038/s41586-022-04593-5. PMID: 35418683

Gurtner GC, Werner S, Barrandon Y, Longaker MT (2008) Wound repair and regeneration. Nature 453:314–321

Horvath S, Singh K, Raj K, Khairnar SI, Sanghavi A, Shrivastava A, Zoller JA, Li CZ, Herenu CB, Canatelli-Mallat M, Lehmann M, Habazin S, Novokmet M, Vučković F, Solberg Woods LC, Martinez AG, Wang T, Chiavellini P, Levine AJ, Chen H, Brooke RT, Gordevicius J, Lauc G, Goya RG, Katcher HL (2024) Reversal of biological age in multiple rat organs by young porcine plasma fraction. Geroscience 46(1):367–394. https://doi.org/10.1007/s11357-023-00980-6. PMID: 37875652

Iram T, Kern F, Kaur A, Myneni S, Morningstar AR, Shin H, Garcia MA, Yerra L, Palovics R, Yang AC, Hahn O, Lu N, Shuken SR, Haney MS, Lehallier B, Iyer M, Luo J, Zetterberg H, Keller

A, Zuchero JB, Wyss-Coray T (2022) Young CSF restores oligodendrogenesis and memory in aged mice via Fgf17. Nature 605(7910):509–515. https://doi.org/10.1038/s41586-022-04722-0.Erratum.In:Nature.2023;613(7942):E1. PMID: 35545674

Izquierdo JM (2024) Blood platelet factor 4: the elixir of brain rejuvenation. Mol Neurodegener 19(1):3. https://doi.org/10.1186/s13024-023-00681-w. PMID: 38185628

Jo H, Brito S, Kwak BM, Park S, Lee MG, Bin BH (2021) Applications of mesenchymal stem cells in skin regeneration and rejuvenation. Int J Mol Sci 22(5):2410. https://doi.org/10.3390/ijms22052410. PMID: 33673711

Katsimpardi L, Litterman NK, Schein PA, Miller CM, Loffredo FS, Wojtkiewicz GR, Chen JW, Lee RT, Wagers AJ, Rubin LL (2014) Vascular and neurogenic rejuvenation of the aging mouse brain by young systemic factors. Science 344(6184):630–634. https://doi.org/10.1126/science.1251141. PMID: 24797482

Kerepesi C, Gladyshev VN (2023) Intersection clock reveals a rejuvenation event during human embryogenesis. Aging Cell 22(10):e13922. https://doi.org/10.1111/acel.13922. PMID: 37786333

Kerepesi C, Zhang B, Lee SG, Trapp A, Gladyshev VN (2021) Epigenetic clocks reveal a rejuvenation event during embryogenesis followed by aging. Sci Adv 7(26):eabg6082. https://doi.org/10.1126/sciadv.abg6082. PMID: 34172448

Kim D, Kiprov DD, Luellen C, Lieb M, Liu C, Watanabe E, Mei X, Cassaleto K, Kramer J, Conboy MJ, Conboy IM (2022) Old plasma dilution reduces human biological age: a clinical study. Geroscience 44(6):2701–2720. https://doi.org/10.1007/s11357-022-00645-w.Erratum.In:Geroscience.2023Nov6; PMID: 35999337

Kiprov DD, Herskowitz A, Kim D, Lieb M, Liu C, Watanabe E, Hoffman JC, Rohe R, Conboy MJ, Conboy IM (2021) Case report: therapeutic and immunomodulatory effects of plasmapheresis in long-haul COVID. F1000Res 10:1189. https://doi.org/10.12688/f1000research.74534.2. PMID: 35464182

Knyazer A, Bunu G, Toren D, Mracica TB, Segev Y, Wolfson M, Muradian KK, Tacutu R, Fraifeld VE (2021) Small molecules for cell reprogramming: a systems biology analysis. Aging (Albany NY) 13(24):25739–25762. https://doi.org/10.18632/aging.203791. PMID: 34919532

Leiter O, Brici D, Fletcher SJ, Yong XLH, Widagdo J, Matigian N, Schroer AB, Bieri G, Blackmore DG, Bartlett PF, Anggono V, Villeda SA, Walker TL (2023) Platelet-derived exerkine CXCL4/platelet factor 4 rejuvenates hippocampal neurogenesis and restores cognitive function in aged mice. Nat Commun 14(1):4375. https://doi.org/10.1038/s41467-023-39873-9. PMID: 37587147

Liu A, Guo E, Yang J, Yang Y, Liu S, Jiang X, Hu Q, Dirsch O, Dahmen U, Zhang C, Gewirtz DA, Fang H (2018) Young plasma reverses age-dependent alterations in hepatic function through the restoration of autophagy. Aging Cell 17(1):e12708. https://doi.org/10.1111/acel.12708. PMID: 29210183

Loffredo FS, Steinhauser ML, Jay SM, Gannon J, Pancoast JR, Yalamanchi P, Sinha M, Dall'Osso C, Khong D, Shadrach JL, Miller CM, Singer BS, Stewart A, Psychogios N, Gerszten RE, Hartigan AJ, Kim MJ, Serwold T, Wagers AJ, Lee RT (2013) Growth differentiation factor 11 is a circulating factor that reverses age-related cardiac hypertrophy. Cell 153(4):828–839. https://doi.org/10.1016/j.cell.2013.04.015. PMID: 23663781

Ma S, Wang S, Ye Y, Ren J, Chen R, Li W, Li J, Zhao L, Zhao Q, Sun G, Jing Y, Zuo Y, Xiong M, Yang Y, Wang Q, Lei J, Sun S, Long X, Song M, Yu S, Chan P, Wang J, Zhou Q, Belmonte JCI, Qu J, Zhang W, Liu GH (2022) Heterochronic parabiosis induces stem cell revitalization and systemic rejuvenation across aged tissues. Cell Stem Cell 29(6):990-1005.e10. https://doi.org/10.1016/j.stem.2022.04.017. PMID: 35613617

McCay CM, Pope F, Lunsford W, Sperling G, Sambhavaphol P (1957) Parabiosis between old and young rats. Gerontologia 1(1):7–17. https://doi.org/10.1159/000210677. PMID: 13405201

Mehdipour M, Skinner C, Wong N, Lieb M, Liu C, Etienne J, Kato C, Kiprov D, Conboy MJ, Conboy IM (2020) Rejuvenation of three germ layers tissues by exchanging old blood plasma with saline-albumin. Aging (Albany NY) 12(10):8790–8819. https://doi.org/10.18632/aging.103418. PMID: 32474458

Mehdipour M, Mehdipour T, Skinner CM, Wong N, Liu C, Chen CC, Jeon OH, Zuo Y, Conboy MJ, Conboy IM (2021) Plasma dilution improves cognition and attenuates neuroinflammation in old mice. Geroscience. 43(1):1–18. https://doi.org/10.1007/s11357-020-00297-8. PMID: 33191466

Muñoz-Cánoves P, Neves J, Sousa-Victor P (2020) Understanding muscle regenerative decline with aging: new approaches to bring back youthfulness to aged stem cells. FEBS J 287(3):406–416. https://doi.org/10.1111/febs.15182. PMID: 31854082

Murach KA, Dimet-Wiley AL, Wen Y, Brightwell CR, Latham CM, Dungan CM, Fry CS, Watowich SJ (2022) Late-life exercise mitigates skeletal muscle epigenetic aging. Aging Cell 21(1):e13527. https://doi.org/10.1111/acel.13527. PMID: 34932867

Muradian KK, Fraifeld VE (2020) IPSCs-induced cellular reprogramming. In: Rattan S (ed) Encyclopedia of biomedical gerontology, vol 2. Elsevier, Academic Press, pp 291–298

Muradian KK, Tolstun DA, Paier AG, Popa-Wagner A, Fraifeld VE (2019) Embryonic stem cells, telomeres and aging. J Ageing Restor Med (JARM) 2(3):115–122. www.scitcentral.com

Muter J, Lynch VJ, McCoy RC, Brosens JJ (2023) Human embryo implantation. Development 150(10):dev201507. https://doi.org/10.1242/dev.201507. PMID: 37254877

Mendelsohn AR, Larrick JW (2013) Rejuvenation of aging hearts. Rejuvenation Res 16(4):330–332. https://doi.org/10.1089/rej.2013.1462. PMID: 23829663

Navarro Negredo P, Yeo RW, Brunet A (2020) Aging and rejuvenation of neural stem cells and their niches. Cell Stem Cell 27(2):202–223. https://doi.org/10.1016/j.stem.2020.07.002. PMID: 32726579

Nguyen N, Sussman MA (2015) Rejuvenating the senescent heart. Curr Opin Cardiol 30(3):235–239. https://doi.org/10.1097/HCO.0000000000000161. PMID: 25760821

Norevik CS, Huuha AM, Røsbjørgen RN, Hildegard Bergersen L, Jacobsen K, Miguel-Dos-Santos R, Ryan L, Skender B, Moreira JBN, Kobro-Flatmoen A, Witter MP, Scrimgeour N, Tari AR (2024) Exercised blood plasma promotes hippocampal neurogenesis in the Alzheimer's disease rat brain. J Sport Health Sci 13(2):245–255. https://doi.org/10.1016/j.jshs.2023.07.003. PMID: 37500010

Pálovics R, Keller A, Schaum N, Tan W, Fehlmann T, Borja M, Kern F, Bonanno L, Calcuttawala K, Webber J, McGeever A, Tabula Muris Consortium, Luo J, Pisco AO, Karkanias J, Neff NF, Darmanis S, Quake SR, Wyss-Coray T (2022) Molecular hallmarks of heterochronic parabiosis at single-cell resolution. Nature 603(7900):309–314. https://doi.org/10.1038/s41586-022-04461-2. PMID: 35236985

Puri D, Wagner W (2023) Epigenetic rejuvenation by partial reprogramming. BioEssays 45(4):e2200208. https://doi.org/10.1002/bies.202200208. PMID: 36871150

Rebo J, Mehdipour M, Gathwala R, Causey K, Liu Y, Conboy MJ, Conboy IM (2016) A single heterochronic blood exchange reveals rapid inhibition of multiple tissues by old blood. Nat Commun 7:13363. https://doi.org/10.1038/ncomms13363. PMID: 27874859

Rigamonti AE, Bollati V, Favero C, Albetti B, Caroli D, Abbruzzese L, Cella SG, Sartorio A (2022) Effect of a 3-week multidisciplinary body weight reduction program on the epigenetic age acceleration in obese adults. J Clin Med 11(16):4677. https://doi.org/10.3390/jcm11164677. PMID: 36012914

Schroer AB, Ventura PB, Sucharov J, Misra R, Chui MKK, Bieri G, Horowitz AM, Smith LK, Encabo K, Tenggara I, Couthouis J, Gross JD, Chan JM, Luke A, Villeda SA (2023) Platelet factors attenuate inflammation and rescue cognition in ageing. Nature 620(7976):1071–1079. https://doi.org/10.1038/s41586-023-06436-3. PMID: 37587343

Sha SJ, Deutsch GK, Tian L, Richardson K, Coburn M, Gaudioso JL, Marcal T, Solomon E, Boumis A, Bet A, Mennes M, van Oort E, Beckmann CF, Braithwaite SP, Jackson S, Nikolich K, Stephens D, Kerchner GA, Wyss-Coray T (2019) Safety, tolerability, and feasibility of young plasma infusion in the plasma for Alzheimer symptom amelioration study: a randomized clinical trial. JAMA Neurol 76(1):35–40. https://doi.org/10.1001/jamaneurol.2018.3288. PMID: 30383097

Sheldrake AR (2022) Cellular senescence, rejuvenation and potential immortality. Proc Biol Sci 289(1970):20212434. https://doi.org/10.1098/rspb.2021.2434. Epub 2022 Mar 2. PMID: 35232226; PMCID: PMC8889192

Shin SH, Lee YH, Rho NK, Park KY (2023) Skin aging from mechanisms to interventions: focusing on dermal aging. Front Physiol 14:1195272. https://doi.org/10.3389/fphys.2023.1195272. PMID: 37234413

Shytikov D, Balva O, Debonneuil E, Glukhovskiy P, Pishel I (2014) Aged mice repeatedly injected with plasma from young mice: a survival study. Biores Open Access 3(5):226–232. https://doi.org/10.1089/biores.2014.0043. PMID: 25371859

Sousa-Victor P, Neves J, Muñoz-Cánoves P (2020) Muscle stem cell aging: identifying ways to induce tissue rejuvenation. Mech Ageing Dev 188:111246. https://doi.org/10.1016/j.mad.2020.111246. PMID: 32311419

Tfilin M, Gobshtis N, Fozailoff D, Fraifeld VE, Turgeman G (2023) Polarized Anti-Inflammatory Mesenchymal Stem Cells Increase Hippocampal Neurogenesis and Improve Cognitive Function in Aged Mice. Int J Mol Sci. 24(5):4490. https://doi.org/10.3390/ijms24054490. PMID: 36901920

Tripathi SS, Kumar R, Arya JK, Rizvi SI (2021) Plasma from young rats injected into old rats induce antiaging effects. Rejuvenation Res 24(3):206–212. https://doi.org/10.1089/rej.2020.2354. PMID: 33161876

Villeda SA, Plambeck KE, Middeldorp J, Castellano JM, Mosher KI, Luo J, Smith LK, Bieri G, Lin K, Berdnik D, Wabl R, Udeochu J, Wheatley EG, Zou B, Simmons DA, Xie XS, Longo FM, Wyss-Coray T (2014) Young blood reverses age-related impairments in cognitive function and synaptic plasticity in mice. Nat Med 20(6):659–663. https://doi.org/10.1038/nm.3569. PMID: 24793238

Xu P, Xin Y, Zhang Z, Zou X, Xue K, Zhang H, Zhang W, Liu K (2020) Extracellular vesicles from adipose-derived stem cells ameliorate ultraviolet B-induced skin photoaging by attenuating reactive oxygen species production and inflammation. Stem Cell Res Ther 11(1):264. https://doi.org/10.1186/s13287-020-01777-6. PMID: 32611371

Xu F, Tautenhahn HM, Dirsch O, Dahmen U (2021) Modulation of autophagy: a novel, "rejuvenation" strategy for the aging liver. Oxid Med Cell Longev 2021:6611126. https://doi.org/10.1155/2021/6611126. PMID: 33628363

Yanai H, Budovsky A, Tacutu R, Fraifeld VE (2011) Is rate of skin wound healing associated with aging or longevity phenotype? Biogerontology 12(6):591–597. https://doi.org/10.1007/s10522-011-9343-6. PMID: 21667230

Yanai H, Toren D, Vierlinger K, Hofner M, Nöhammer C, Chilosi M, Budovsky A, Fraifeld VE (2015) Wound healing and longevity: lessons from long-lived αMUPA mice. Aging (Albany NY) 7(3):167–176. https://doi.org/10.18632/aging.100726.Erratum.In:Aging(Albany NY).2016;8(3):564. PMID: 25960543

Yanai H, Budovsky A, Tacutu R, Barzilay T, Abramovich A, Ziesche R, Fraifeld VE (2016) Tissue repair genes: the TiRe database and its implication for skin wound healing. Oncotarget 7(16):21145–21155. https://doi.org/10.18632/oncotarget.8501. PMID: 27049721

Yang JH, Petty CA, Dixon-McDougall T, Lopez MV, Tyshkovskiy A, Maybury-Lewis S, Tian X, Ibrahim N, Chen Z, Griffin PT, Arnold M, Li J, Martinez OA, Behn A, Rogers-Hammond R, Angeli S, Gladyshev VN, Sinclair DA (2023) Chemically induced reprogramming to reverse cellular aging. Aging (Albany NY) 15(13):5966–5989. https://doi.org/10.18632/aging.204896. PMID: 37437248

Yankova T, Dubiley T, Shytikov D, Pishel I (2022) Three month heterochronic parabiosis has a deleterious effect on the lifespan of young animals, without a positive effect for old animals. Rejuvenation Res 25(4):191–199. https://doi.org/10.1089/rej.2022.0029. PMID: 35747947

Yücel AD, Gladyshev VN (2024) The long and winding road of reprogramming-induced rejuvenation. Nat Commun 15(1):1941. https://doi.org/10.1038/s41467-024-46020-5. PMID: 38431638

Zhang B, Lee DE, Trapp A, Tyshkovskiy A, Lu AT, Bareja A, Kerepesi C, McKay LK, Shindyapina AV, Dmitriev SE, Baht GS, Horvath S, Gladyshev VN, White JP (2023) Multi-omic rejuvenation and life span extension on exposure to youthful circulation. Nat Aging 3(8):948–964. https://doi.org/10.1038/s43587-023-00451-9. PMID: 37500973

Conclusions

Although rejuvenation ideas are so old that nobody knows how old they are, *Rejuvenology* is a new-emerging science with huge perspectives and matching problems. More people are becoming convinced that living younger and longer is not a utopia. After all, the fundamental element of the living system—the cell—is potentially immortal! The main principles and mechanisms of immortality have been fixed in the genome and transferred to the next generations, including cells of metazoans. Immortal prokaryotes and protozoa demonstrate that there are various ways to cellular immortality, alas, they are not used in the vast majority of metazoans. It could be an important reserve for future research.

Why do multicellular organisms, consisting of potentially immortal units, have 'surrendered' to aging so easily? Where are the immortal metazoans? Yet, there are a few exceptions when multicellular species show no signs of aging or only negligible senescence, with life span exceeding hundreds and thousand years.

The last years have been manifested by an impressive increase in efforts to reverse aging by systemic interventions or in a tissue-specific manner. However, it should be noticed that there are a lot of respectful people who do not want to live longer. Religion, traditions, principal considerations, health and social problems, psychology, and many other factors could be the reason. However, the global population enjoys the continuing increase in human life span and would like to live much longer. Among them are many talented and persistent young scientists who make research on rejuvenation a rapidly growing science. Before 1970, the analysis of PubMed publications with the keyword "rejuvenation" showed that there were only 100 scientific articles! Since then, the number of new publications has doubled every 5 years, and currently it has reached over 12,000 papers with nearly 100,000 authors. The 100,000 'motivated' researchers are a serious power to be sure that *Rejuvenology* has been established as a scientific discipline. It has specific methodology and methods, extremely difficult but sensible goals.

Rejuvenology is an unusual science. As an emerging one, it meets numerous challenges. One of them is that many people do not believe in a radical extension of

K. K. Muradian and V. E. Fraifeld, *Rejuvenation and Longevity*, Healthy Ageing and Longevity 20, https://doi.org/10.1007/978-3-031-64995-0

life span and rejuvenation. It is quite expectable, because from the very childhood one gets used to the idea that aging and death are unavoidable realities, while rejuvenation and immortality are for fairy tales. It is, however, a delusion.

Human knowledge and instrumental power increase *exponentially* whereas the complexity of aging and rejuvenation problems is a *constant* entity. This is a good reason to believe that, sooner or later, human skills will be enough for resolving any biomedical problem, anti-aging and rejuvenation included. In these coordinates, there is only one unknown—when? Will it take 30 years, 50 years, 100 years or 300 years? Maybe it occurred yesterday, you just missed the latest news? Anyway, active dispute and robust expectations definitely indicate that systemic rejuvenation of a multicellular organism is on the agenda. To reach the summits, a hard climb is yet ahead. Welcome, rejuvenation!

Index

K. K. Muradian and V. E. Fraifeld, *Rejuvenation and Longevity*, Healthy Ageing and Longevity 20, https://doi.org/10.1007/978-3-031-64995-0